Essentials of Radio Wave Propagation

If you need to maximise efficiency in wireless network planning an understanding of radio propagation issues is vital, and this quick reference guide is for you. Using real-world case studies, practical problems and minimum mathematics, the author explains simply and clearly how to predict signal strengths in a variety of situations.

Fundamentals are explained in the context of their practical significance. Applications, including point-to-point radio links, broadcasting and earth–space communications, are thoroughly treated, and more sophisticated methods, which form the basis of software tools both for network planning and for spectrum management, are also described.

For a rapid understanding of and insight into radio propagation, sufficient to enable you to undertake real-world engineering tasks, this concise book is an invaluable resource for network planners, hardware designers, spectrum managers, senior technical managers and policy makers who are either new to radio propagation or need a quick reference guide.

CHRISTOPHER HASLETT is the Principal Propagation Adviser at Ofcom, the UK Communication Industries Regulator. As well as experience conducting and directing research projects, he has many years' industrial radio-planning experience with Cable and Wireless plc., and as Director of Planning and Optimisation at Aircom International Ltd., where he directed the optimisation of UMTS networks. He was also a Senior Lecturer at the University of Glamorgan.

T0332253

The Cambridge Wireless Essentials Series

Series Editors
WILLIAM WEBB, *Ofcom, UK*
SUDHIR DIXIT, *Nokia, US*

A series of concise, practical guides for wireless industry professionals.

Martin Cave, Chris Doyle and William Webb, *Essentials of Modern Spectrum Management*
Christopher Haslett, *Essentials of Radio Wave Propagation*

Forthcoming
Andy Wilton and Tim Charity *Essentials of Wireless Network Deployment*
Steve Methley *Essentials of Wireless Mesh Networking*
Malcolm Macleod and Ian Proudler *Essentials of Smart Antennas and MIMO*
Stephen Wood and Roberto Aiello *Essentials of Ultra-Wideband*
David Crawford *Essentials of Mobile Television*
Chris Cox *Essentials of UMTS*

For further information on any of these titles, the series itself and ordering information see www.cambridge.org/wirelessessentials.

Essentials of Radio
Wave Propagation

Christopher Haslett
Ofcom, UK

CAMBRIDGE
UNIVERSITY PRESS

CAMBRIDGE UNIVERSITY PRESS
Cambridge, New York, Melbourne, Madrid, Cape Town,
Singapore, São Paulo, Delhi, Mexico City

Cambridge University Press
The Edinburgh Building, Cambridge CB2 8RU, UK

Published in the United States of America by Cambridge University Press, New York

www.cambridge.org
Information on this title: www.cambridge.org/9780521875653

First published 2008

A catalogue record for this publication is available from the British Library

ISBN 978-0-521-87565-3 Hardback

Contents

Preface

The objective of this book is to allow the reader to predict the received signal power produced by a particular radio transmitter. The first two chapters examine propagation in free space for point-to-point and point-to-area transmission, respectively. This is combined with a discussion regarding the characteristics of antennas for various purposes. In chapter 3, the effect of obstacles, whether buildings or mountains, is discussed and analytical methods, whereby the strength of a signal is the shadow of an obstacle can be predicted, are presented. The following chapter investigates the nature of reflections and the effect that reflections have on the nature of a received signal. Chapter 5 shows how the level of a received signal can be predicted considering all propagation mechanisms. The many effects on a radio wave that are caused by precipitation and the structure of the atmosphere are explained in chapter 6. Chapter 7 demonstrates how knowledge gained can be used to design point-to-point radio links, broadcast systems, Earth–space systems and in-building systems. In chapter 8, the value of software tools in the planning of various radio networks is explained.

Acknowledgements

I believe that understanding of a topic such as radio wave propagation seeps in over the years rather than being something that can be accumulated rapidly. This is always facilitated by discussion and debate with others in the field. For such discussion and debate I am particularly indebted to David Bacon, Les Barclay, Ken Craig, Martin Hall, Malcolm Hamer, Tim Hewitt, Mireille Levy, John Pahl, James Richardson, Richard Rudd, Mike Willis and Charles Wrench.

1 Propagation in free space and the aperture antenna

This chapter introduces the basic concepts of radio signals travelling from one antenna to another. The aperture antenna is used initially to illustrate this, being the easiest concept to understand. The vital equations that underpin the day-to-day lives of propagation engineers are introduced. Although this chapter is introductory in nature, practical examples are covered. The approach adopted is to deliver the material, together with the most significant equations, in a simplified manner in the first two subsections before providing more detail. Following this, the focus is on developing methods of predicting the received signal power on point-to-point links given vital information such as path length, frequency, antenna sizes and transmit power.

1.1 Propagation in free space: simplified explanation

Radio waves travel from a source into the surrounding space at the 'speed of light' (approximately 3.0×10^8 metres per second) when in 'free space'. Literally, 'free space' should mean a vacuum, but clear air is a good approximation to this. We are interested in the power that can be transmitted from one antenna to another. Because there are lots of different antennas, it is necessary to define a reference with which others can be compared. The isotropic antenna in which the transmitted power is radiated equally in all directions is commonly used as a reference. It is possible to determine the ratio between the power received and that transmitted in linear units, but it is more common to quote it in decibels (dB). Further information on the decibel scale is given in an appendix at the back of this book. If we have an isotropic antenna as transmitter and receiver then the loss in dB is given by

$$\text{loss} = 32.4 + 20 \log d + 20 \log f, \tag{1.1}$$

where d is the path length in kilometres and f is the frequency in MHz. This loss in free space between two reference, isotropic, antennas is known as the 'free-space loss' or 'basic transmission loss'. The difference between the transmitted power and received power on any point-to-point system (the 'link loss') is the free-space loss less any antenna gains plus any miscellaneous losses. To maintain consistency, any antenna gains must be quoted relative to the reference isotropic antenna. Again it is normal to use the decibel scale and the gain is quoted in 'dBi', with the 'i' indicating that we are using an isotropic reference antenna:

$$\text{link loss} = 32.4 + 20\log d + 20\log f - G_t - G_r + L_m, \qquad (1.2)$$

where G_t and G_r are the gains of the transmitting and receiving antennas, respectively, in dBi and L_m represents any miscellaneous losses in the system (such as feeder or connector losses) in dB.

1.2 The aperture antenna: simplified explanation

The antenna forms the interface between the 'guided wave' (for example in a coaxial cable) and the electromagnetic wave propagating in free space. Antennas act in a similar manner irrespective of whether they are functioning as transmitters or receivers, and it is possible for an antenna to transmit and receive simultaneously. The simplest form of antenna to visualise is the aperture antenna. Parabolic dishes used for microwave communications or satellite Earth stations are good examples of aperture antennas. The gain of an aperture antenna increases with increasing antenna size and also increases with frequency. The gain of a circular parabolic dish type of aperture antenna is given by the approximation

$$\text{gain (dBi)} \approx 18 + 20\log D + 20\log f, \qquad (1.3)$$

where D is the diameter of the dish in metres and f is the frequency of operation in GHz. Thus the gain of an antenna will increase by 6 dB if it doubles in diameter or if the frequency of operation doubles. Aperture antennas must be accurately pointed because they generally have narrow beamwidths. The beamwidth is usually measured in degrees. A useful approximation is

$$\text{beamwidth} \approx \frac{22}{Df} \text{ degrees}, \qquad (1.4)$$

where D is the diameter of the dish in metres and f is the frequency of operation in GHz. So, as the diameter increases, the beamwidth gets smaller and, similarly, as the frequency increases the beamwidth gets smaller.

The antenna is a reciprocal device, acting equally well as a transmitter as it does as a receiver. A high-quality transmitting antenna will radiate almost all the power entering its feed into the surrounding space. In a lower-quality antenna, a significant amount of this power will be dissipated as heat or reflected back along the feeder to the transmitter. An antenna that reflects a lot of the power back along the feeder is said to be 'poorly matched'.

1.3 Further details and calculations

Assuming that the transmitting antenna is perfect (that is, all the power entering from the feeder cable is radiated into space) makes the calculations simpler. Using linear units, the power entering the antenna from its feed, P_t, is measured in watts. Once it has left the antenna, it creates a power density, P_d, in space that is measured in watts per square metre. One very useful concept in radio wave propagation studies is known as the 'isotropic antenna'. This (fictional) antenna is thought of as radiating power equally in all directions. It is straightforward to predict the power density created by an isotropic antenna at a certain distance. To clarify this, we need to conduct a 'thought experiment'. Consider an isotropic antenna located at the centre of a sphere. Since the antenna radiates equally in all directions, the power density must be equal on all parts of the surface of the sphere. Further, since the sphere encloses the antenna, all of the power radiated by the antenna must pass through this sphere. Combining these two pieces of information, and knowing that the area of a sphere of radius r metres is equal to $4\pi r^2$ square metres, allows us to write the equation

$$P_d = \frac{P_t}{4\pi r^2}. \qquad (1.5)$$

This equation reveals a very valuable generalisation in radio wave propagation: the 'inverse square' law. It can be seen that the power density produced by an antenna reduces with the square of the distance. This is true in free space and any deviation from this in 'real-world' environments is a matter of interest to radio-wave scientists.

We now turn our attention to the antenna as a receiver. It is a very common situation for a single antenna to transmit and receive simultaneously. Thus any practical antenna will possess attributes appropriate for transmission and reception of radio waves. When receiving, an antenna is illuminated by a particular power density and has the job of converting this into a received power in its feeder cable. Again a thought experiment is helpful. Consider a radio wave of a particular power density travelling through space. Now consider an aperture such that any power entering is passed into the feeder cable. The power entering the aperture (the received power, P_r) depends on the size of the aperture, A_e square metres (the suffix e taken to stand for 'effective' when referring to a receiving antenna), and the power density of the radio wave:

$$P_r = P_d A_e. \tag{1.6}$$

From our equation linking P_d to P_t we can now write

$$P_r = \frac{P_t A_e}{4\pi r^2}. \tag{1.7}$$

This gives us the power received at distance r metres by an antenna with effective aperture A_e square metres when an isotropic antenna transmits power P_t watts.

At this point we must turn our attention once more to the transmitting antenna. So far, we have considered this antenna to transmit power equally in all directions (isotropically). Practical antennas do not do this. The fact that antennas possess directivity also leads to the phenomenon known as antenna gain.

When most engineers think about a device possessing 'gain' they assume that the device outputs more signal power than it takes in. Antennas, as passive devices, cannot do this. The best they can do is to radiate all the power they receive from the feeder. When we talk about the gain of an antenna, we refer to the fact that the directivity leads to the power being concentrated in particular directions. Thus, if you enclosed the antenna in a sphere, centred at the antenna, you would not see equal power densities at all points on the sphere. The point on the sphere where the power density is greatest indicates the 'principal direction' of the antenna. The power density is given more generally by

$$P_d = \frac{P_t G_t}{4\pi r^2},$$ (1.8)

where G_t is the gain of the transmitting antenna in any direction. The principal direction, that of maximum gain (the direction in which the antenna is pointing), is the most usual direction to consider. Using our modified equation for the power density, we can modify the equation for the received power:

$$P_r = \frac{P_t G_t A_{er}}{4\pi r^2},$$ (1.9)

where the effective aperture of the receiving antenna is now called A_{er} to make it clear that it is the receiving antenna that is being referred to.

Remember that the same antenna will transmit and receive. Thus the same antenna will possess gain as a transmitting antenna and an effective aperture as a receiving antenna. There is a link between the two, which we shall now investigate.

In the thought experiment that follows it is easiest if you allow yourself to imagine the antenna as a parabolic dish type of antenna such as those used for microwave communications or for satellite Earth stations. Such antennas are in fact known as 'aperture antennas' because their 'effective aperture' is very easy to picture; the reflecting dish itself is the aperture. It should be noted that wire antennas (such as those used for terrestrial television reception and on mobile devices) possess an effective aperture, even though this is less obvious from visual examination of the antenna. When an aperture antenna is used to transmit, power is directed towards the parabolic reflector from a feed placed at the focus. This power is then reflected back into space. Aperture antennas are known for producing narrow beams. The 'beamwidth' of an aperture antenna is influenced by the two following rules: the bigger the parabolic dish, the narrower the beam; the higher the frequency, the narrower the beam. It is this ability to focus, or collimate (collimate means 'form into a column' and describes

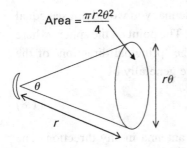

Figure 1.1 The area illuminated by an aperture antenna with a perfectly conical beam.

the way the energy travels without much spreading), the energy into a narrow beam that endows the antenna with gain.

Consider an idealised aperture antenna that radiates the energy in a perfectly conical beam of angle θ radians. If θ is small, a circle of diameter $r\theta$ would be evenly illuminated at distance r. The area of this circle is $\pi r^2 \theta^2 / 4$ (see figure 1.1). Now, an isotropic antenna would illuminate a sphere of area $4\pi r^2$. The aperture antenna illuminates a smaller area. The same amount of power is concentrated into a smaller area. The reduction in area illuminated equals the increase in power density. It is this increase in power density that we call the gain of the antenna. Thus the gain of the antenna in this case is related to its beamwidth:

$$gain = \frac{4\pi r^2}{\pi r^2 \theta^2 / 4}$$
$$= 16/\theta^2. \tag{1.10}$$

Remember that the beamwidth has been idealised into a perfect cone shape and is not therefore exact for a practical antenna. It is not, however, a bad approximation. Thus, if we are told the beamwidth of an antenna, we can estimate its gain. It is more common to be told the beamwidth of the antenna in degrees.

Example: an antenna has a beamwidth of 3 degrees. Estimate its gain.

Answer:

$$3 \text{ degrees} = 3 \times \pi \div 180 = 0.052 \text{ radians};$$
$$gain = 16/0.052^2 = 5800.$$

It is common to express the gain in logarithmic units (decibels):

$$\text{gain in dBi} = 10 \log_{10}(5800) = 37.7 \,\text{dBi}.$$

The additional suffix 'i' indicates that the gain has been calculated with respect to an isotropic antenna.

Thus we can estimate that the gain of an antenna with a beamwidth of 3 degrees is 37.7 dBi. However, a further important question is 'How big would this antenna be?'. In order to answer this question, we need to accept a crucial fact about the isotropic antenna. We need to know the effective aperture of an isotropic antenna. Throughout this book, equations are derived in as straightforward a manner as possible. However, the derivation of the effective aperture of an isotropic antenna, though crucial, is beyond the scope of this book and we must accept it as the proven work of others. The effective aperture of an isotropic antenna, A_{ei}, depends on the wavelength, λ, and is given by

$$A_{ei} = \frac{\lambda^2}{4\pi}. \tag{1.11}$$

This is actually the area of a circle whose circumference equals λ.

It is more common to talk about the frequency of operation than the wavelength. The wavelength in metres can be converted into the frequency f in MHz by use of the equation

$$\lambda = \frac{300}{f(\text{MHz})} \text{ metres}. \tag{1.12}$$

Thus, if we use the units of square metres for the effective aperture, the effective aperture of an isotropic antenna is given by

$$A_{ei} = \frac{300^2}{4\pi f^2} = \frac{7160}{f^2} \text{ square metres}. \tag{1.13}$$

The gain of an antenna is numerically equal to its effective aperture expressed as a multiple of the effective aperture of an isotropic antenna at the frequency of interest. Therefore, knowledge of the effective aperture of an isotropic antenna at a particular frequency is crucial when it comes to determining the gain of an antenna.

A further thought experiment is necessary to clarify a vital feature about antennas: the radiation pattern, and hence directivity and gain, is exactly the same when the antenna is transmitting as when it is receiving. In order to convince yourself of this, imagine an antenna placed in a sealed chamber. Suppose that this antenna is connected to a matched load. This load both receives thermal noise gathered from the antenna and transfers thermal noise to the antenna. The same amount of power is transferred in both directions, thus maintaining equilibrium. If the radiation pattern as a transmitter differed from that as a receiver, the antenna would transfer energy from one part of the chamber to another in violation of the laws of thermodynamics. This fact means that it does not matter whether you consider an antenna as a transmitter or as a receiver; the radiation pattern is the same. If we can calculate its gain as a receiver, then we know its gain as a transmitter.

Let us consider the antenna that we previously calculated to have a gain of 5800. If it is to have this gain as a receiver, the effective aperture must be 5800 times that of an isotropic antenna. The actual aperture size in square metres depends on the frequency. At 5000 MHz, the effective aperture of an isotropic antenna is 2.86×10^{-4} square metres. An antenna with a gain of 5800 would have an effective aperture of 1.66 square metres. If it was a circular aperture, its diameter would be 1.45 metres. This indicates the size of parabolic dish that would be required in order to have a gain of 5800 (and a beamwidth of 3 degrees) at a frequency of 5 GHz. In fact, this would be for a perfect aperture antenna. Practical antennas have a smaller aperture than that calculated from the diameter of the dish. This reduction in aperture is accounted for by the term 'aperture efficiency', η, which lies between 0 and 1. The actual size of a circular reflecting parabolic dish of diameter D is $\pi D^2/4$. Thus, considering aperture efficiency, the effective aperture of any practical circular parabolic dish is given by

$$A_e = \eta \frac{\pi D^2}{4}, \qquad (1.14)$$

where D is the diameter of the parabolic dish. This is illustrated in figure 1.2.

If we repeated the calculations at a frequency of 10 GHz, we would find that the diameter of the antenna would be halved for the same gain and beamwidth. This is one main reason why microwave communication

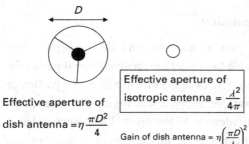

Figure 1.2 The effective aperture of an isotropic antenna and a practical aperture antenna.

Effective aperture of isotropic antenna $= \dfrac{\lambda^2}{4\pi}$

Effective aperture of dish antenna $= \eta \dfrac{\pi D^2}{4}$

Gain of dish antenna $= \eta \left(\dfrac{\pi D}{\lambda}\right)^2$

has become popular: the higher the frequency, the easier it is to collimate the energy. Some general rules follow.

- If you double the frequency, the gain of an antenna will quadruple.
- If you double the frequency, the beamwidth of an antenna will halve.
- If you double the antenna diameter (keeping the frequency the same), the gain of the antenna will quadruple.
- If you double the antenna diameter (keeping the frequency the same), the beamwidth of an antenna will halve.

In our example, we took an antenna whose beamwidth was 3 degrees. At a frequency of 5 GHz, the required diameter was found to be 1.45 metres. If we multiply the frequency (in GHz) by the diameter (in metres) and by the beamwidth (in degrees) we find $5 \times 1.45 \times 3 \approx 22$. Because of the way in which beamwidth (BW) halves if frequency doubles and so on, we find that this formula is a good general approximation:

$$\text{BW (degrees)} \times D\,(\text{m}) \times f\,(\text{GHz}) \approx 22. \qquad (1.15)$$

The fact that the beamwidth reduces as the antenna diameter increases poses significant challenges to the radio-system engineer. The beamwidth gives an indication of how accurately the antenna must be pointed. Antennas on masts are subject to wind forces that will tend to affect the antenna direction. The larger the antenna, the larger the wind force that acts on it. The fact that larger antennas must also be pointed more accurately and cannot therefore be allowed to alter direction by as much as smaller antennas can leads to the necessity for careful consideration before larger antennas are selected.

1.4 Point-to-point transmission

With knowledge about the gain of antennas and the free-space loss between two points it is possible to predict the received signal power for a particular situation and, thence, to design a link to a deliver a particular power to the receiver. For example, suppose that we have a system consisting of two parabolic dish antennas 15 km apart. Each antenna has a diameter of 1 m. Ignoring any feeder or miscellaneous losses, we can estimate the link loss at a frequency of, say, 20 GHz. The gain of the antennas will each be approximately $18 + 20\log(1) + 20\log(20) = 44$ dBi. Therefore the link loss will be given by $32.4 + 20\log(15) + 20\log(20\,000) - 44 - 44 = 53.9$ dB.

Thus, if a received signal power of -40 dBm were required, a transmit power of 13.9 dBm would be needed. If we repeat the exercise for similar-sized antennas at a higher frequency, the loss is found to be less. It must be remembered that this is a prediction for propagation in a vacuum. Although the loss will be nearly as low as this for much of the time in the Earth's atmosphere, atmospheric effects will cause 'fading', leading to the need for a substantial 'fade margin'.

It is possible to do a quick estimate for the necessary receive power in a microwave point-to-point system. It depends upon the bit rate. The required power is approximately $-154 + 10\log(\text{bit rate})$ dBm. Microwave links usually demand high performance and low errors. This in turn means high power. Mobile systems typically operate at a higher bit error rate and can operate at lower power levels, perhaps 10 dB lower than for microwave links of the same bit rate.

1.4.1 Transmitting, directing and capturing power

We now have a picture of an antenna, both in transmit and in receive 'mode'. As a transmitter, the antenna directs energy into space. As a receiver, the antenna presents an aperture to the incident electromagnetic wave in order to gather energy. Our understanding can be enhanced by examining a point-to-point link in a different way. Let us consider our parabolic dish antenna with a perfect, conical, radiation pattern and a beamwidth of 3 degrees. At a distance of 1 kilometre, the antenna would illuminate a circle with a diameter of approximately 50 metres. It would in principle be possible to establish a receiving antenna of diameter 50 metres at this distance and gather nearly all of the transmitted energy (in practice no antenna has a perfectly conical radiation pattern, but it would be possible to capture say 80% of the transmitted energy). In this way point-to-point radio transmission in free space can be designed to be almost lossless. You will, however, hear radio engineers talking of 'free-space loss' and yet the term itself is something of an oxymoron, a contradiction in terms: free space is a lossless medium. What engineers mean when they talk about free-space loss is the difference between transmit and receive powers between two standard antennas (the 'standard' antenna usually being an isotropic antenna). This loss is less commonly (but more accurately) referred to as 'spreading loss', thus indicating that the energy spreads out as it travels further from the transmitting antenna. It would be highly unusual, to say the least, actually to install a 50-metre-diameter receiving antenna. Suppose, in fact, that the receiving antenna has the more realistic diameter of only 1 metre. This is a fiftieth of the diameter required to capture all of the power (1/2500 of the area). This would capture approximately 1/2500 of the transmitted power. Thus the free-space loss, or spreading loss, would be 2500 or, in dB, $10 \log_{10}(2500) = 34.0$ dB. This is illustrated in figure 1.3.

Let us now use the same thought process to determine the transmission loss between two isotropic antennas.

At a distance of r metres, the area illuminated by an isotropic transmitting antenna would be $4\pi r^2$ square metres. The effective aperture of an isotropic antenna is given by equation (1.13) as $7160/f^2$ (MHz) square

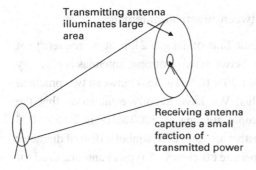

Transmitting antenna illuminates large area

Receiving antenna captures a small fraction of transmitted power

Figure 1.3 An illustration of spreading loss on a point-to-point link.

metres. Thus the free-space transmission loss, L_{fs}, between two isotropic antennas is given by

$$L_{fs} = \frac{4\pi r^2}{7160/f^2 \,(\text{MHz})} = 0.001\,76(rf)^2. \tag{1.16}$$

This equation is more commonly seen in a slightly different form. Firstly, it is more normal to think of distances in kilometres rather than metres. This gives

$$L_{fs} = 1760(fd)^2, \tag{1.17}$$

where d is the distance in kilometres and f is the frequency in MHz. Secondly, it is more common to quote free-space loss in logarithmic units, decibels (dB):

$$10\log_{10}(L_{fs}) = 10\log_{10}1760 + 10\log_{10}(f^2) + 10\log_{10}(d^2),$$
$$L_{fs}\,(\text{dB}) = 32.4 + 20\log_{10}(f) + 20\log_{10}(d)\,\text{dB}. \tag{1.18}$$

The above equation is well-remembered by radio-propagation engineers. Remember that this is the transmission loss between two isotropic antennas. Although isotropic antennas do not actually exist, the above equation is very important because it provides a reference from which to compare real radio links. Before we move on to this, it is worth pointing out that you will see a variation in the form of the above equation. We have been using MHz as the unit for frequency. An equivalent form of the equation exists, using GHz as the frequency. Since $1000\,\text{MHz} = 1\,\text{GHz}$ and $20\log_{10}(1000) = 60$, the equation becomes

$$L_{fs}\,(\text{dB}) = 92.4 + 20\log_{10}(f\,(\text{GHz})) + 20\log_{10}(d)\,\text{dB}. \tag{1.19}$$

1.4.2 Transmission loss between practical antennas

Let us consider a point-to-point link of length 2 km at a frequency of 5000 MHz. The free-space loss between two isotropic antennas is given by $1760(fd)^2 = 1760(5000 \times 2)^2 = 1.76 \times 10^{11}$. The loss between two practical antennas will be less than this. We have already established that the effective aperture of an isotropic antenna at 5000 MHz is 2.86×10^{-4} square metres and that the effective aperture of a parabolic dish of diameter D is $\eta \pi D^2/4$, where η is the aperture efficiency. A typical antenna used for point-to-point links has an efficiency of about 60% ($\eta = 0.6$). Thus an antenna with a (typical) diameter of 1.2 metres would have an effective aperture of $0.6 \times \pi \times 1.2^2 \times 0.25 = 0.68$ square metres. This means that the antenna would have a gain of $0.68/(2.86 \times 10^{-4}) = 2400$. Now, in a point-to-point link there would be two such antennas. Thus the overall transmission loss would be reduced to $1.76 \times 10^{11}/(2400 \times 2400) = 31000$. Thus, if we wish to receive 1 microwatt of power we would have to transmit at 31 000 microwatts (or 31 milliwatts). The receiving antenna 'captures' only 1/31 000 of the transmitted power. The vast majority of the transmitted power misses the receiving antenna and continues out into space, its power density weakening as the energy disperses.

Although the fact that 1/31 000 of the transmitted power ends up being received at the far end of the link gives us a realistic picture of what is happening, calculations are generally performed using logarithmic (decibel) units. The gain of the antenna would be described not as 2400 but as 33.8 dBi ($10 \log_{10} 2400 = 33.8$). The free-space loss would be calculated in dB as $32.4 + 20 \log_{10} 5000 + 20 \log_{10} 2 = 112.4$ dB. The overall loss would then be calculated as $112.4 - 33.8 - 33.8 = 44.8$ dB. This equals 31 000 expressed in decibels ($10 \log_{10} 31\,000 = 45$). The situation is illustrated in figure 1.4.

Thus, if we transmit at a power of, say, +10 dBm, we will receive a signal power of −34.8 dBm.

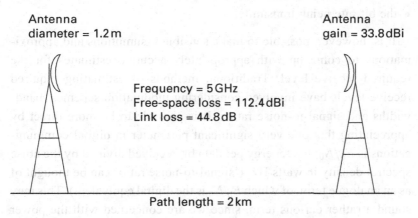

Antenna
diameter = 1.2 m

Antenna
gain = 33.8 dBi

Frequency = 5 GHz
Free-space loss = 112.4 dBi
Link loss = 44.8 dB

Path length = 2 km

Figure 1.4 Significant parameters on a practical microwave link.

One last approximation before we sum up: we calculated the antenna gain as a ratio of its effective aperture to that of the effective aperture of an isotropic antenna at 5000 MHz (=5 GHz). In order to calculate the transmission loss in decibels, we needed to convert the gain into dB. It is possible to do this directly using the following approximation:

$$\text{antenna gain (dBi)} \approx 18 + 20 \log_{10} f \text{ (GHz)} \\ + 20 \log_{10} D \text{ (metres)}. \qquad (1.20)$$

Using this approximation, we could immediately estimate the gain of a 1.2-metre-diameter parabolic dish to be 33.6 dBi at a frequency of 5 GHz, which is very close to our previous estimate (any differences being due to different assumptions regarding aperture efficiency).

1.4.3 Determining the power required

When designing point-to-point systems, it is common to design to a minimum required receive signal power. It is useful to have an idea of the powers that are generally required. The power required by any radio receiver depends on a number of things:

- the quality of the receiver,
- the noise and interference being received,
- the required bit error ratio,
- the modulation scheme used and
- the bit rate being transmitted.

It is, however, possible to make sensible assumptions and approximations to come up with appropriately accurate estimates for the required receive level. Traditional methods of estimating required receive levels have involved considering modulation schemes, bandwidths and signal-to-noise ratios. It is possible to be more direct by appreciating that one very significant parameter in digital communications is E_b/N_0: the 'energy per data bit received divided by the noise spectral density in watts/Hz' ('signal-to-noise ratio' can be thought of as an analogue term, of which E_b/N_0 is the digital equivalent). This may sound a rather curious term, since we are concerned with the power required to maintain a steady stream of bits, not just one bit. However, if we stay with the argument for a while, a useful result emerges. The actual value of N_0 depends on the quality of the particular receiver. N_0 can be written as generally equal to kT, where k is Boltzmann's constant of 1.38×10^{-23} joules per kelvin and T is the noise temperature in kelvins. The noise gathered by an antenna depends on the temperature of whatever the antenna is 'looking at'. On Earth, in a terrestrial point-to-point system, the antenna is 'looking at' the Earth, buildings, vegetation etc. and the atmosphere. The noise temperature cannot be less than the aggregate temperature of these artefacts. This is often given the standard temperature of 290 kelvins (290 K), or approximately 17 °C. Thus, a reference value of N_0 for terrestrial systems is $1.38 \times 10^{-23} \times 290 = 4.0 \times 10^{-21}$ joules. In order to achieve an acceptable bit-error ratio, a practical receiver will require energy per bit of about 100 times this amount, or 4.0×10^{-19} joules. Now we need to convert this figure into a required receive power. To do this, we simply multiply by the bit rate. So, a microwave backhaul link at 8 Mbit/s will require a minimum receive signal power of approximately $4.0 \times 10^{-19} \times 8 \times 10^6 = 3.2 \times 10^{-12}$ watts, or -85 dBm. This minimum required signal level is known as the 'threshold' level. If the received signal power decreases below this level then the bit-error ratio will become unacceptably high. Thus we have arrived at an important rule of thumb:

$$\text{required receive power in a terrestrial system}$$
$$\approx 4.0 \times 10^{-19} \times \text{bitrate (watts)}$$
$$= -154 + 10 \log (\text{bitrate}) \text{ dBm.} \qquad (1.21)$$

Some important points to note are the following.

- This figure has been based on an assumed level of thermal noise. Interference will add to this. In cases where the receiver is subject to significant interference, the required receive power will increase.
- The value of the noise temperature, T, has been assumed to be 290 kelvins. This is an estimate of a typical atmospheric temperature. It is largely influenced by the temperature of whatever the antenna is looking at. Satellite Earth-station receivers are pointed at free

space, which is very cold (approximately 4 kelvins). In most Earth–space systems, the path includes some intervening atmosphere that will lead to an increase of this value but, for a well-designed satellite Earth station, a value for T of 60 kelvins would be more usual.

- This figure makes an assumption regarding the level of acceptable bit-error ratio. The assumption made here is reasonably valid for point-to-point microwave systems where the expectation is of high quality. Where the link is for a mobile radio system using hand-portable terminals, users will accept a lower quality and a lower receive energy of perhaps 4.0×10^{-20} joules per bit will be acceptable.

1.4.4 Example calculations

We now know enough about antennas (particularly aperture antennas) and propagation through free space to experiment with some examples to see what results we get.

Question: A point-to-point system operates over a distance of 20 kilometres at a frequency of 26 GHz. The antennas are each of diameter 90 cm. Estimate the beamwidths of the antennas deployed and the power received if the transmit power is 20 dBm.

Solution: Estimating beamwidth from

$$\text{beamwidth} \times \text{diameter} \times \text{frequency} \approx 22,$$
$$\text{beamwidth} \times 0.9 \times 26 \approx 22,$$
$$\text{beamwidth} \approx 22/(0.9 \times 26)$$
$$\approx 0.94 \text{ degrees.}$$

To estimate the power received, first determine the free-space loss between isotropic antennas over that distance at that frequency:

$$\text{free-space loss} = 92.4 + 20 \, \log(d) + 20 \, \log(f)$$
$$= 92.4 + 20 \, \log(20) + 20 \, \log(26)$$
$$= 146.7 \text{ dBi.}$$

Estimate the gain of the antennas from

$$\text{gain} \approx 18 + 20 \ \log(f) + 20 \ \log(D)$$
$$\approx 18 + 20 \ \log(26) + 20 \ \log(0.9)$$
$$\approx 45.4 \ \text{dBi}.$$

Thus the overall loss is the free-space loss between isotropic antennas less the gain of the antennas deployed. Remember, we use two antennas (one at each end). Thus the net loss is $146.7 - 45.4 - 45.4 = 55.9 \ \text{dB}$. If the power transmitted is 20 dBm, then the power received will be $20 - 55.9 = -35.9 \ \text{dBm}$.

Question: A geostationary satellite is 39 000 km from an Earth station. It is transmitting a digital signal at a bit rate of 36 Mbit/s using a 40-dBm transmitter at a frequency of 11.2 GHz. The transmitting antenna has a diameter of 80 cm. Determine the required size of a receiving antenna if the required E_b/N_0 ratio is 12 dB and the noise temperature of the receiving system is 160 kelvin. Boltzmann's constant, $k = 1.38 \times 10^{-23}$.

Solution: Note that 12 dB is a logarithmic ratio equal to $10^{1.2}$ in linear units. $N_0 = kT = 1.38 \times 10^{-23} \times 160 = 2.21 \times 10^{-21}$ joules. Required $E_b = 2.21 \times 10^{-21} \times 10^{1.2} = 3.50 \times 10^{-20}$ joules. The required receive power is $3.50 \times 10^{-20} \times 36 \times 10^6 = 1.26 \times 10^{-12}$ watts. This equals $10 \ \log(1.26 \times 10^{-12}) = -119.0 \ \text{dBW} = -89.0 \ \text{dBm}$. Thus the maximum link loss that can be tolerated is $40.0 - (-89.0) = 129.0 \ \text{dB}$.

The free-space loss (FSL) at a frequency of 11.2 GHz and a distance of 39 000 km is

$$92.4 + 20 \ \log(11.2) + 20 \ \log(39 \ 000) = 205.2 \ \text{dBi}.$$

The gain of the transmitting antenna can be estimated from its diameter (80 cm) and knowing the operating frequency of 11.2 GHz:

$$G_t = 18.0 + 20 \ \log(11.2) + 20 \ \log(0.8) = 37.0 \ \text{dBi}.$$

Now, ignoring miscellaneous losses,

$$\text{link loss} = \text{FSL} - (G_t + G_r).$$

For the maximum allowed link loss of 129.0 dB,

$$129.0 = 205.2 - (37.0 + G_r),$$
$$G_r = 205.2 - 129.0 - 37.0$$
$$= 39.2.$$

Thus the receiving antenna must have a gain of 39.2 dBi. The diameter of such an antenna can be estimated from the equation

$$gain = 18.0 + 20\log(f) + 20\ \log(D)$$
$$39.2 = 18.0 + 20\log(11.2) + 20\log(D)$$
$$20\ \log(D) = 39.2 - 18.0 - 20\ \log(11.2) = 0.2$$
$$\log(D) = 0.01$$
$$D = 1.0 \text{ metres.}$$

1.4.5 Equivalent isotropic radiated power (EIRP)

When considering radio links, we are often concerned with determining the power received for a particular configuration.

We know that the power received is given by $P_r = P_t - \text{link loss} = P_t + G_t + G_r - \text{FSL}$.

It is common to regard $P_t + G_t$ as a single entity because it is the sum of the two terms that is significant rather than the individual terms of transmitted power and transmitting antenna gain; $P_t + G_t$ is known as the 'equivalent isotropic radiated power' and is given the abbreviation EIRP. The same power would be received if a real power equal to the EIRP were transmitted through an isotropic antenna as would be received through the actual configuration. Thus the same power would be received if a power of 50 dBm were transmitted through an antenna of 15 dBi gain as for a power of 40 dBm transmitted through an antenna of 25 dBi gain. The EIRP in both cases is 65 dBm.

1.5 Near-field effects: simplified explanation

The equations quoted so far make assumptions regarding the link that we are dealing with. The free-space-loss equation has a '20 log(distance)' term. As the distance becomes small, this can become hugely negative, resulting in the possibility of negative link loss being predicted for very short links (receive power bigger than transmit power!). This is because the equations are valid only in the 'far field' of the antennas. The extent of the 'near-field' region (within which the equations are not valid) depends upon the frequency and the size of the antenna. There is no exact boundary between the near and the far field but a common rule is to use the expression ($fD^2/150$) as the extent of the near field, where f is the frequency in MHz and D is the antenna diameter in metres. If you know the gain of the antenna in dBi but information on the antenna diameter is not to hand, an alternative approximation can be used:

$$\text{near-field distance} \approx \frac{100 \times 10^{\text{gain}/10}}{f\,(\text{MHz})} \text{ metres.} \qquad (1.22)$$

A 1-metre antenna operating at a frequency of 20 GHz (20 000 MHz) would have a near-field distance of approximately 130 metres. Such an antenna would have a gain of about 44 dBi. Using the equation involving the gain of the antenna produces a similar estimate for the near-field distance. The equations for free-space loss and link loss can be used only for distances greater than the near-field distance. Since microwave links are designed for communications over distances upwards of a few kilometres, this is not usually a problem.

1.5.1 Further detail on the near-field issue

We have developed some simple equations that will allow us to calculate the receive signal power given details of the frequency, path length, transmit power and the transmit and receive antennas. As an example, let us assume that a particular transmitter delivers a power of +20 dBm into the feeder of the transmitting antenna. We are operating at a frequency of 30 GHz over a distance of 12 km. The transmitting and receiving antennas are of 0.9 metres diameter.

We can estimate the gain of the two antennas from

$$\text{antenna gain (dBi)} \approx 18 + 20 \log_{10} f \,(\text{GHz}) + 20 \log_{10} D \,(\text{metres})$$
$$= 18 + 20 \log_{10} 30 + 20 \log_{10} 0.9$$
$$= 46.6 \,\text{dBi.}$$

Thus the combined antenna gains would be $2 \times 46.6 = 93.2$ dBi.

The free-space loss is given by

$$L_{fs} \text{ (dB)} = 92.4 + 20 \log_{10}(f \text{ (GHz)}) + 20 \log_{10}(d) \text{ dB}$$
$$= 92.4 + 20 \log_{10}(30) + 20 \log_{10}(12) \text{ dB}$$
$$= 143.5 \text{ dB}.$$

Thus the overall transmission loss is $143.5 - 93.2 = 50.3$ dB. If the transmitted power is 20 dBm, then the received power is $20 - 50.3 = -30.3$ dBm.

The loss of 50.3 dB is approximately equal to a ratio of $1 : 100\,000$. That is, only $1/100\,000$ of the power transmitted is captured by the receiving antenna. Remember that the reason for the loss is not that there is any lossy medium present but, rather, that the transmitting antenna could not prevent the radiated energy dispersing as it travelled towards the receiver. At the distance of the receiver, the aperture of the receiving antenna could capture only this tiny fraction of the power that was transmitted.

Suppose now that we perform a similar calculation but this time we will make the antennas bigger and the path length shorter. We shall consider the situation where the antennas are 5 metres in diameter and the distance between the transmit and receive antennas is only 100 metres. Now the gain of the antennas is estimated by

$$\text{antenna gain (dBi)} \approx 18 + 20 \log_{10} 30 + 20 \log_{10} 5 = 61.5 \text{ dBi}$$

and the free-space loss is now

$$92.4 + 20 \log_{10}(30) + 20 \log_{10}(0.1) \text{ dB} = 101.9 \text{ dB}.$$

This means that the transmission loss would be calculated as $101.9 - (2 \times 61.5) = -21.1$ dB. This negative value of loss suggests that, if the transmission power were $+20$ dBm then the receive power would be $+41.1$ dBm. This is clearly nonsense; it is impossible to receive more power than is transmitted. We must look at our assumptions in an attempt to determine what has gone wrong. One equation that we used in deriving a method for determining the transmission loss linked the power received, P_r, with the incident power density, P_d, is $P_r = P_d A_e$, where A_e is the effective aperture of the receiving antenna. One implicit assumption is that the power density is uniform over the surface of the receiving aperture. A quick calculation suggests that the beamwidth of the transmitting antenna (frequency 30 GHz, diameter 5 m) is approximately $22/(30 \times 5) = 0.15°$. At a distance of 100 metres, this beamwidth would illuminate a circle of only 25 cm diameter, whereas the receiving aperture has a diameter of 5 metres. We are clearly in a situation where our simplified equations are not valid. This mental picture of only a tiny portion of the receiving aperture being illuminated is not true but the conclusion that our equations are not valid at such short distances for such large antennas is most certainly true.

Another assumption necessary in order to make the equations valid is that the transmitting antenna can be regarded as a point source when viewed from the receiver (we regarded the idealised antenna as having a conical beam, a cone converges to a point). An antenna of 5 m diameter viewed from a distance of 100 metres is clearly not a point source. To illustrate this, let us consider the manner in which the reflecting parabolic dish helps to form the transmitted beam. The dish is illuminated by a small feeder. The feeder itself is a form of aperture antenna but with a wide beamwidth (sufficient to illuminate the majority of the reflecting dish). The parabolic dish then reflects the power in a forward direction; the bigger the dish the better the beam-forming properties (the narrower beamwidth leading to less dispersion). This method imposes a maximum power density that can be produced: it is approximately the total power transmitted divided by the surface area of the dish ('approximately' because the feeder will not illuminate the dish evenly). Thus the maximum power density that could be produced by a dish of 5 metres diameter transmitting at a power of 100 mW (20 dBm = 100 mW = 0.1 watts) $= 0.1 \times 4/(\pi \times 5^2) = 5.1 \times 10^{-3}$ watts per square metre. However, if we calculate the power density that would be predicted using the equation $P_d = P_t G_t/(4\pi r^2)$ (firstly converting the antenna gain to linear units: the estimated antenna gain of 61.5 dBi $= 1.4 \times 10^6$) we get $0.1 \times 1.4 \times 10^6/(4\pi \times 100^2) = 1.1$ watts per square metre. This predicted value is massively above the maximum that can be produced by this antenna. We are violating our point-source assumption. Of course, the point-source assumption is never 'true' – the dish of diameter 5 metres is never a 'point'. However, the further we are away from the transmitter, the smaller the error introduced by making this approximation becomes. Radio engineers would say that, at only 100 metres away, we are in the 'near field' of the antenna. How far does the near field extend? It depends on the diameter of the antenna and the frequency of operation.

If we look at the calculations made for the power density produced by the transmitting antenna, we would feel more comfortable if the predicted power density were much less than the maximum power density that we know is possible. A common interpretation of 'much less' is 'less than one tenth', so, if the predicted field strength were less than 5.1×10^{-4} watts per square metre, we would feel more comfortable. The distance for which this power density would be predicted is

$$\sqrt{\frac{(0.1 \times 1.4 \times 10^6)}{\pi \times 5.1 \times 10^{-4}}} = 9300 \text{ metres}.$$

A commonly used approximation for the 'near-field distance' (NFD) is $2D^2/\lambda$, where D is the diameter of the dish and λ is the wavelength of transmission (a frequency of 30 GHz has a wavelength of 0.01 m). Thus, a 5-metre-diameter dish would have an estimated near-field distance of 5000 metres at a frequency of 30 GHz. At this distance, the predicted power density would be approximately one third of the maximum possible. The higher the frequency, the larger the near-field distance. There is no absolute border on going from the near field to the far field. It is a gradual transition. It is often more convenient to consider the frequency of operation rather than the wavelength. If the frequency, f, is in MHz, it is related to wavelength by the equation $\lambda = 300/f$ metres. Thus the near-field distance, NFD, equates to

$$\text{NFD} = \frac{fD^2}{150} \text{ metres}. \tag{1.23}$$

Now, the gain of a parabolic dish antenna is given by

$$\text{gain} = \frac{\eta \pi D^2/4}{7160/f^2} \tag{1.24}$$

but we know that NFD $= fD^2/150$ and hence

$$D^2 = \frac{150(\text{NFD})}{f}. \tag{1.25}$$

Therefore,

$$\text{gain} = \frac{150\eta\pi(\text{NFD})/(4f)}{7160/f^2} \tag{1.26}$$

and hence

$$\text{NFD} = \frac{\text{gain} \times 4 \times 7160}{150\eta\pi f}. \tag{1.27}$$

Using a typical approximation for the value of the aperture efficiency, η, of 0.6 gives an approximate value for the near-field distance of

$$\text{NFD} \approx \frac{\text{gain} \times 100}{f \text{ (MHz)}}. \tag{1.28}$$

Note that the gain in the above equation is in linear units, not in dBi. If the gain is quoted in dBi, then

$$\text{NFD} \approx \frac{100 \times 10^{\text{gain(dBi)}/10}}{f \text{ (MHz)}} \text{ metres.} \tag{1.29}$$

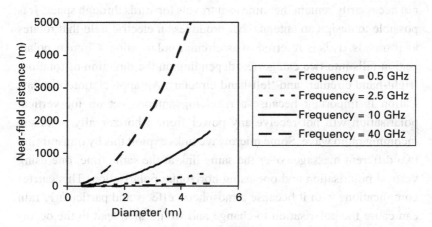

Figure 1.5 Graphs showing how the near-field distance of antennas varies with diameter.

This gives us a way of estimating the near-field distance when only the gain of the antenna is known, together with the frequency of operation. This shows that, for a given gain, the NFD decreases with increasing frequency whereas, for a given antenna diameter, the NFD increases with increasing frequency. The graph in figure 1.5 shows how the NFD varies with diameter for a range of frequencies.

1.6 Polarisation

Radio waves are electromagnetic waves. As the name suggests, they consist of an electric field and a magnetic field. These waves travel through space together, keeping in phase with each other. The directions of the electric and magnetic fields are at right angles to the direction of propagation (electromagnetic waves are 'transverse' waves) and to each other. By convention, the direction of the electric field determines what is known as the polarisation of the radio wave. All the equations that we have so far used for determining the loss on a link assume that the polarisation of the transmitting antenna is the same as that of the receiving antenna. The most common polarisations in use for microwave links are vertical and horizontal. A vertical half-wave dipole antenna will radiate with vertical polarisation. The direction of the electric field does not necessarily remain the same as it travels forwards through space. It is possible to design an antenna that produces an electric field that rotates as it travels. This is referred to as circular polarisation. Circular polarisation falls into two categories, depending on the direction of rotation: 'right-hand circular' and 'left-hand circular'. An appreciation of polarisation is important because a receiving antenna set up for vertical polarisation will not receive any power from a horizontally polarised incoming radio wave. Some microwave links exploit this by transmitting two different messages over the same link at the same time; one using vertical polarisation and one using horizontal polarisation. This carries complications with it because atmospheric effects and particularly rain can cause the polarisation to change and would then lead to the occurrence of interference between the two polarisations. The polarisation of a radio wave can rotate as it propagates. This is noticeable on Earth–space links at lower microwave frequencies (below about 10 GHz) when the rotation occurs as the wave propagates through the ionosphere. One advantage of circular polarisation is that rotation does not affect it: it remains circular. For this reason, circular polarisation is commonly used in links to geostationary satellites at frequencies below 10 GHz.

If a vertically polarised wave reflects off a surface that is not vertical or horizontal, its polarisation will be changed (this effect is similar to the way in which objects viewed via reflection in a mirror can appear at a different orientation if the angle of the mirror is varied). This means that it is difficult for the polarisation of waves that can undergo many reflections (such as in mobile communications) to be predicted accurately.

1.7 Summary

The manner in which a radio wave propagates in free space has been introduced. It is almost impossible to present a separate discussion on this subject without involving the role played by the antenna at each end of a radio link. Antennas form the vital interface between the physical equipment involved in a radio link and free space. It has been shown that aperture antennas such as microwave dishes produce a beam that can be described by its beamwidth in degrees. Particular attention has been paid to determining the overall loss on point-to-point links and equations to enable this have been explained. Limitations to the equations have been examined, with particular attention being paid to the near-field distance. Finally, the significance of polarisation in radio transmission has been explained.

2   Point-to-area transmission

Point-to-area transmission is the generic name given to the way in which broadcasting transmitters or base stations for mobile communications provide coverage to a given area. A general overview is provided in order to deliver the most significant information as quickly as possible. Following that, the concept of electric field strength as an alternative to power density, prediction methods and the effect of frequency are explained in more detail. Digital mobile radio is selected for further study as a specific example of point-to-area communication. Path-loss-prediction methods specific to digital mobile radio are examined and various types of antenna that are used for the base stations of these systems are described. Finally, the effect that interference can have on the coverage range of a base station is explained.

2.1 Overview

The simplest form of antenna used is an 'omni-directional' antenna. These radiate equally in all directions in the horizontal plane. They often have a narrower beam (perhaps less than 20 degrees) in the vertical plane. Such antennas typically have a gain of 10 dBi. Two collinear wire elements that are fed at their junction with a signal form what is known as a dipole antenna. The basic omni-directional antenna is a dipole that is half a wavelength in height (a 'half-wave dipole' with each of the wire elements being a quarter of a wavelength in length). This has a wide vertical beam and has a gain of 2.1 dBi. It is used as a reference antenna in a similar manner to an isotropic antenna and gains of broadcast-type antennas are often quoted in dB relative to a half-wave dipole, given the symbol dBd (so a gain of 10 dBi is the same as a gain of 7.9 dBd).

Because the half-wave dipole is more commonly used as the reference than an isotropic antenna in these situations, the term EIRP is replaced by

ERP ('effective radiated power'). The ERP is the term $P_t + G_t$ (in logarithmic units), where the gain of the antenna is expressed in dBd. Thus an ERP of 50 dBm equals an EIRP of 52.1 dBm.

A typical omni-directional antenna used in a base station consists of an array of perhaps eight half-wave dipoles (or similar) stacked in a straight vertical line. An alternative antenna popular in mobile communications is the 'sectored' antenna. This has a horizontal beamwidth of about 80 degrees and covers a region of about 120 degrees in angle (the signal does not just reduce to zero straight away once you move out of the nominal beamwidth). This consists of an omni-directional antenna plus a reflector to provide the required coverage pattern. These have a higher gain than do omni-directional antennas (about 16 dBi).

For purposes such as television broadcasting, high-power transmitters are fed to antennas in elevated positions. The receiver typically uses a directional antenna mounted at rooftop height providing additional gain to the transmission link. Typical domestic television antennas have a gain of about 14 dBi. This means that a single transmitter can provide coverage over a 100 kilometres or more, even when broadcasting television pictures for which a large bandwidth is required. Antennas for mobile communications are designed for coverage of smaller areas. The fact that the mobile terminal has an omni-directional antenna (with gain of about 0 dBi) and that this terminal is not at rooftop height but often only a metre or so above ground height surrounded by buildings ('down in the clutter') means that the link losses are higher than for broadcast systems. Base stations for mobile communications can provide coverage up to a few tens of kilometres in rural areas but, in urban areas, the coverage range is restricted to a kilometre or so, particularly if communication to mobile terminals located indoors is to be provided.

When it is required to predict the coverage from a broadcasting transmitter, it must be acknowledged that the terrain will vary with direction of transmission. If it can be assumed that the broadcasting antenna is placed well above any surrounding 'clutter' (such as nearby buildings) then it is possible to use a 'path-general' prediction model ('path-general' models do not consider every nuance of the path profile between the transmitter and receiver) to obtain a rapid prediction of the power density produced at points in the surrounding area. Recommendation ITU-R P. 1546 contains a path-general model that is widely used in predicting power densities produced by broadcasting transmitters.

2.2 Power density and electric field strength

Point-to-area transmission refers to transmission and/or reception by one central antenna to/from other transmitters or receivers spread throughout what is known as its coverage area. The two major types of point-to-area transmission are broadcasting (radio and television) and digital mobile communications (such as GSM and UMTS). Television broadcasting predictions are carried out assuming that the receiving antenna is placed above the roof and also has a directional antenna that is pointed at the base station. Television reception does, however, require a large signal level because of its wide bandwidth (around 8 MHz for analogue television). A domestic receiver would need a signal of approximately -80 dBm in order to deliver a clear picture.

We have so far spoken of the power density produced by a transmitting antenna. An alternative, which is commonly used when investigating broadcast antennas, is to consider the electric field strength. There is a direct link between power density and electric field strength. The energy in a radio signal travels as an electromagnetic wave. The main components are, unsurprisingly, an electric field (E volts per metre) and a magnetic field (H amperes per metre). The power density is given by the product of the two fields:

$$P_d = EH \text{ watts per square metre.} \qquad (2.1)$$

Note that in many books you will see power density, more correctly, depicted as the vector cross-product of the two fields ($\hat{P} = E \times H$), which also indicates the direction of travel of the electromagnetic wave. However, we are just interested in finding out the magnitude of the power density, and then simply multiplying the electric field strength by the magnetic field strength will give the correct answer. When considering radio waves, the magnetic field gets little mention. When wire antennas are used as receivers, the picture we imagine is of the electric field inducing currents in the antenna. In free space (and, as far as we are concerned, in air) the ratio of the electric and magnetic fields in an electromagnetic wave is constant (the ratio E/H in free space is called the 'intrinsic impedance' of free space).

In free space,

$$E = 120\pi H \tag{2.2}$$

and hence the power density can be determined in terms of the electric field strength alone:

$$P_d = \frac{E^2}{120\pi}. \tag{2.3}$$

Electric field strength is often preferred to power density in performing calculations with broadcast transmitters. Transforming the above equation gives

$$E = \sqrt{P_d 120\pi}$$
$$= \sqrt{\frac{P_t G_t}{4\pi r^2} 120\pi}$$
$$= \frac{\sqrt{30 P_t G_t}}{r}, \tag{2.4}$$

where E is measured in volts per metre and r is measured in metres. Because received electric field strengths are very small, the microvolt per metre has become a standard measure. Similarly, distances are more commonly measured in kilometres. Therefore

$$E \ (\mu V/m) = 1\,000\,000 \frac{\sqrt{30 P_t G_t}}{1000 d} = 5480 \frac{\sqrt{P_t G_t}}{d}. \tag{2.5}$$

In a further piece of standardisation or simplification, it is common to take a value of $P_t G_t$ as 1000 watts transmitted through a half-wave dipole (a half-wave dipole has a gain of 2.14 dBi, which is 1.64 as a ratio). Thus a standard value of $P_t G_t$ is 1640. In this way

$$E \ (\mu V/m) = 5480 \frac{\sqrt{1640}}{d} = \frac{222\,000}{d}. \tag{2.6}$$

As always, as a final manoeuvre, we usually express this in logarithmic units (dB relative to 1 microvolt/m, or dBμV/m):

$$E_{dBμV/m} = 20 \log \left(\frac{222\,000}{d} \right) = 106.9 - 20 \log d. \qquad (2.7)$$

Notice that this equation, unlike the equation for free-space loss, has no frequency term, making it easier to use in certain circumstances, particularly when the sensitivity of a receiving system is expressed in dBμV/m (as is often the case with receivers of broadcast signals) rather than dBm.

2.3 Converting from field strength to received signal power

If we know the electric field strength at the receiver, we can determine the power density. This can be converted to power received by multiplying the power density by the effective aperture of the receiving antenna. If the receiving antenna is unknown, then the power that would be received by an isotropic antenna at that location is often predicted:

$$P_r(\text{isotropic}) = \frac{E^2}{120\pi} \frac{7160}{f^2(\text{MHz})} \text{ watts}, \qquad (2.8)$$

where E is the field strength in volts per metre.

If E is expressed in microvolts per metre and P_r is measured in milliwatts, then the equation becomes

$$P_r \text{ (milliwatts)} = \frac{E(μV/m)^2}{1×10^9 × 120\,\pi f^2 \text{ (MHz)}} 7160$$

$$= 1.9×10^{-8} \frac{E\,(μV/m)^2}{f^2\,(\text{MHz})}. \qquad (2.9)$$

Converting to logarithmic units gives

$$P_r \text{ (dBm)} = -77.2 + 20 \log E \text{ (μV/m)} - 20 \log f \text{ (MHz)}. \qquad (2.10)$$

Finally, on noting that

$$20 \log E \text{ (μV/m)} = E \text{ (dBμV/m)} \qquad (2.11)$$

we get

$$P_r \text{ (dBm)} = E \text{ (dBμV/m)} - 20 \log f \text{ (MHz)} - 77.2. \qquad (2.12)$$

Remember that this is the power that would be received by an isotropic antenna. The power received by a directional antenna could be obtained by adding the gain (in dBi) to this value.

2.4 Predicting the field strength at a distance

The reason why we have spent so much time appreciating the significance of field strength is that the recommendation ITU-R P. 1546 provides predictions of field strength. It does this in the form of families of curves at various frequencies, heights of base station and time percentages (when considering interference, the strongest signal experienced for only 1% of the time is significant because interference can be described as harmful even if it exists only for short periods of time). As an example let us investigate the curve (shown in figure 2.1) that predicts the signal exceeded for 50% of the time (the 'median' level) at a frequency of 600 MHz when the transmitting antenna is 75 metres high and transmits with an 'effective radiated power', ERP, of 1000 W (the effective radiated power is the

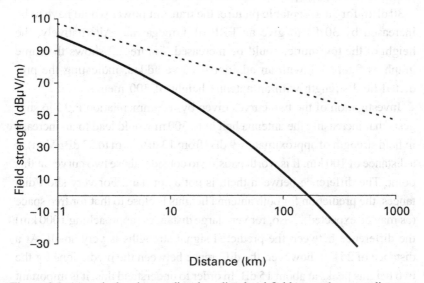

Figure 2.1 A graph showing predicted median-level field strength versus distance for a base-station antenna of height 75 m transmitting with an effective radiated power of 1000 watts at a frequency of 600 MHz. The dashed line shows the prediction assuming free-space propagation ($106.9 - 20 \log d$).

product of the actual power and the gain of the antenna relative to a half-wave dipole – thus the curves are intended to predict the electric field strength when the actual power is 1000 W and a half-wave dipole is used or when a lower power is transmitted and compensated for by a higher-gain antenna).

The graph shows, for example, that the field strength at a distance of 100 km is predicted to be approximately 13 dBμV/m. We can convert this to a power that would be received by an isotropic antenna at a frequency of 600 MHz using the equation

$$P_r \text{ (dBm)} = E \text{ (dBμV/m)} - 20 \log f \text{ (MHz)} - 77.2$$
$$= 13 - 20 \log(600) - 77.2$$
$$= -119.7. \tag{2.13}$$

A domestic television receiving aerial has a gain of about 14 dBi and would thus deliver approximately −105 dBm into the feeder cable. There would be further losses between the antenna and the television, reducing it to perhaps −110 dBm. In order to deliver the necessary −80 dBm for an acceptable picture, the transmit power would have to be increased by 30 dB to give an ERP of 1 megawatt. Alternatively, the height of the transmitter could be increased. Figure 2.2 shows the same graph as figure 2.1 with an additional curve added, indicating the predicted field strength with an antenna height of 300 metres.

Investigation of the two curves given by recommendation P. 1546 suggests that increasing the antenna height to 300 m would lead to an increase in field strength of approximately 9 dB (from 13 dBμV/m to 22 dBμV/m) at a distance of 100 km. It is worth pausing to consider these two curves at this point. The difference between them is not a constant. For very short distances, the prediction for both antenna heights is close to that for free space (as may be expected). Also, for very large distances (approaching 1000 km) the difference between the predicted signal strengths is very small. At a distance of 25 km, however, the difference between the predictions for the two heights peaks at about 15 dB. In order to understand this, it is important to appreciate the effect of the curvature of the Earth.

When you look out at the horizon, the distance to the horizon depends on your height. Let's assume that you have a view of the sea, in which

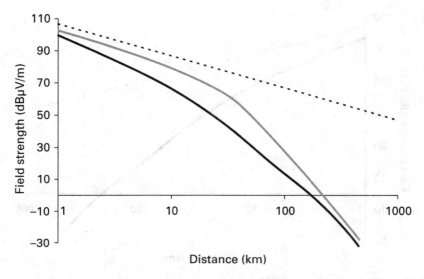

Figure 2.2 A graph showing predicted median-level field strength versus distance for base-station antennas of heights 75 m (solid line) and 300 m (faded line) transmitting with an effective radiated power of 1000 watts at a frequency of 600 MHz. The dashed line shows the prediction assuming free-space propagation.

case the distance to the horizon depends upon your height above sea level. The distance to the visible horizon is approximately $3.6\sqrt{h}$ kilometres, where h is the height above sea level in metres. The radio horizon, for reasons we shall discuss later, is somewhat larger than the visible horizon, approximately $4.1\sqrt{h}$ kilometres. Thus, at a height of 75 metres, the distance to the radio horizon is 35 km, whereas at a height of 300 metres it is 70 km. The difference between the two predictions is greatest for distances just less than and just greater than the radio horizon for a height of 75 metres. Once the distance exceeds 70 km we are 'over the horizon' for both heights and the difference decreases. At 1000 km we are well over the horizon and the height of the transmitting antenna has little influence on the received signal level.

2.5 The effect of frequency

One useful feature of using electric field strength as the indicator at a distance rather than power density is that the free-space equation then

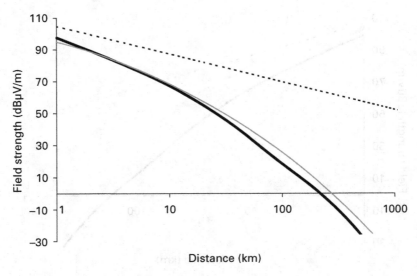

Figure 2.3 Predictions of median level for an antenna height of 75 m at frequencies of 600 MHz (solid line) and 100 MHz (faded line).

has no frequency term. The electric field strength under free-space conditions at a distance d km is given by

$$E_{fs} = 106.9 - 20 \log d \text{ dB}\mu\text{V/m}. \tag{2.14}$$

When a graph of electric field strength against distance with distance on a logarithmic scale is drawn, a straight line results. Thus, for short distances where the receiver will have a good view of the transmitter, the electric field strength will be close to the free-space value for all frequencies. However, when the link is longer and the bulge of the Earth starts to have an impact, the field strength will depend on frequency. Figure 2.3 shows the way in which the field strength reduces with distance for frequencies of 100 MHz and 600 MHz. As expected, for short distances (up to about 30 km) there is little discernible difference between the two curves. For larger distances, the field strength is predicted to be greater at the lower frequency. This is due to the fact that electromagnetic waves at lower frequencies tend to diffract around obstacles better than do waves at higher frequencies. In this case the 'obstacle' concerned is the bulge of the Earth.

2.6 Prediction methods for digital mobile communication

The P. 1546 method is recommended for use for frequencies in the range 30–3000 MHz and distances of 1–1000 km. The curves provided by P. 1546 predict the field strength at a height of 10 m above the ground. This corresponds fairly well to the height of a roof of a typical two-storey building. However, mobile terminals are often at street level. This means that not only are they at a lower height than that for which the curves predict, but also they are often in the shadow of buildings. The P. 1546 method does include a height gain factor to consider such situations but designers of mobile networks usually rely on other different propagation models specifically designed for such purposes. Designers do not need a propagation model that is valid for distances up to 1000 km and much of the value of P. 1546 would be negated. Additionally, speed of computation is crucial because many coverage calculations would be needed in order to give details of the field strength produced over a large area.

We shall consider two such models here:

- Okumura–Hata
- Walfisch–Ikegami

2.6.1 The Okumura–Hata model

This model, developed in 1980 by Hata [1] and based on measurements reported by Okumura *et al.* [2] in 1968, can be simplified when used for a particular frequency such as 900 MHz and a typical mobile antenna height of 1.5 metres to give

$$\text{loss} = 146.8 - 13.82 \log h + (44.9 - 6.55 \log h)\log d \text{ dB}, \quad (2.15)$$

where d is the distance from the base-station in kilometres and h is the height of the base station antenna in metres. It is claimed to be valid for ranges of distance from 1 km to 20 km and for ranges of base-station height from 30 metres to 200 metres.

When requiring a quick check on coverage, an engineer may well assume a particular base-station antenna height such as 30 metres.

Then the equation simplifies to

$$loss = 126.4 + 35.2 \log d. \qquad (2.16)$$

A pan-European 'COST' project (COST action 231) entitled 'Digital mobile radio towards future generation systems' further developed each of the two above-mentioned models. Since GSM was beginning to occupy frequencies around 1800 MHz, development of the models to be applicable in the range 1500–2000 MHz was a key goal. This goal was achieved. When planning a network, a planning engineer will need to know the most appropriate model for a particular frequency band. For a typical frequency of 1800 MHz and a standard assumed height of a mobile antenna of 1.5 metres, the model put forward as suitable for an urban environment can be written as

$$loss = 157.3 - 13.82 \log h + (44.9 - 6.55 \log h)\log d, \qquad (2.17)$$

where d is the distance from the base station in kilometres and h is the height of the base-station antenna in metres. It is claimed to be valid for ranges of distance from 1 km to 20 km and for ranges of base station height from 30 metres to 200 metres.

When requiring a quick check on coverage, an engineer may again assume a typical base-station antenna height such as 30 metres. Then the equation simplifies to

$$loss = 136.9 + 35.2 \log d. \qquad (2.18)$$

The minimum signal required by a GSM mobile is approximately −105 dBm.

A 50-watt (47-dBm) transmitter with an antenna gain of 16 dBi (typical for a sectored antenna) would have a transmit EIRP of 63 dBm if there were no feeder losses. Feeder losses would reduce this to approximately 60 dBm and therefore a loss of 165 dB could be tolerated. If an allowance for building penetration of 20 dB is provided for, then a maximum loss of 145 dB must be targeted. The equation must be transposed to determine the maximum range,

$$d = 10^{\text{loss}-136.9}/35.2$$
$$= 10^{145-136.9}/35.2$$
$$= 1.7\,\text{km}, \tag{2.19}$$

suggesting a coverage range of 1.7 km for a typical, high-power GSM base station operating at 1800 MHz with a high-gain sectored antenna at a height of 30 metres. In a later section we will see that an extra margin for error needs to be added to the system losses in order to give us confidence that coverage will be obtained. This entails adding in a margin of about 5 dB, thus reducing the range by a factor of $10^{5/35.2} = 1.4$. This reduces the expected range from a base station from 1.7 km to 1.2 km.

A further model, which is a combination of two separate models by Walfisch and Ikegami (and hence known as the Walfisch–Ikegami model), is more complex than the Okumura–Hata model. It requires knowledge of the street width and orientation of the street relative to the direction of the base station. Such data might not be readily available.

The starting point of the Walfisch–Ikegami method is an equation similar in style to the Okumura–Hata equation, with the exception that terms accounting for 'rooftop-to-street diffraction' and 'multi-screen diffraction' are included. Determining suitable values for these extra terms requires knowledge about roof heights and street widths. A further correction for street orientation is added.

The Walfisch–Ikegami method requires more information to be available. It is also slower to compute. Its advantages are that it has validity (although with reduced accuracy) even when the transmitting antenna is below the surrounding roof height. It is generally felt to be superior to the Okumura–Hata method when predicting signal strength over short distances in an urban environment (the minimum range of validity of the Okumura–Hata method is 1 km; the Walfisch–Ikegami method can be used over distances as short as 20 metres).

2.6.2 Comparison between propagation at 900 MHz and propagation at 1800 MHz

For coverage estimates at a frequency of 900 MHz we have been using the following approximate equation: loss $= 126.4 + 35.2 \log d$. The equation for free-space loss has a 20 log f element that suggests that, if the frequency is doubled to 1800 MHz, then the loss will increase by $20 \log 2 = 6$ dB. Indeed, we have seen that a suitable approximation for estimating path loss at 1800 MHz is loss $= 136.9 + 35.2 \log d$. The loss at 1800 MHz is approximately 10 dB greater than that at 900 MHz. The free-space loss is 6 dB greater at 1800 MHz but, additionally, the radio waves will not propagate as well around buildings at 1800 MHz as they do at 900 MHz. When designing antennas, great attention is paid to the shape of the beam required. Because the shape of the beam and its beamwidth are directly related to the antenna gain, the gains of antennas used in digital mobile communications systems will be approximately the same at 900 MHz as they are at 1800 MHz. The only advantage that using 1800 MHz has is that the antennas will be physically smaller. If antennas of the same gain are used, the range of a 900-MHz antenna will be significantly greater than that of a 1800-MHz antenna. The path-loss difference of 10.5 dB translates into a difference in range of $-10^{10/35.2} = 1.92$. We have previously estimated the range from a base station when providing indoor coverage in an urban environment to be 1.2 km at 1800 MHz. The range at 900 MHz would be estimated to be 2.3 km.

2.7 Base-station antennas

The simplest form of base-station antenna is an omni-directional half-wave dipole. This is a wire antenna and, if mounted with the wire of the antenna vertical, has an omni-directional pattern in the horizontal plane. In the vertical plane it exhibits directivity with nulls in the vertically up and down directions and a maximum in the horizontal direction. It has a gain relative to an isotropic antenna of 2.14 dBi. The half-wave dipole is such a standard antenna that it has become a benchmark in its own right. Many base-station antennas have their gains quoted in the unit 'dBd', the letter d indicating that the half-wave dipole, not the isotropic, is the reference. Remember: 0 dBd $= 2.14$ dBi. Figure 2.4 gives the radiation pattern of a half-wave dipole in the vertical direction (assuming that the elements are arranged vertically).

On some small installations such as those used for private mobile radio systems by taxi operators and the like, a single half-wave dipole

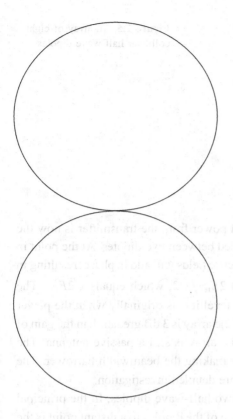

Figure 2.4 The radiation pattern of a half-wave dipole. The principal direction is perpendicular to the line of the wire elements, with nulls in the direction of the elements.

(or similar antenna such as the 'folded dipole') may be used. Antennas used for purposes such as GSM and UMTS communications inevitably use more than one half-wave dipole arranged to form an array. The half-wave dipoles are usually stacked equally spaced in a straight line (a 'collinear array'). An array of eight half-wave dipoles is shown in figure 2.5.

Let us now perform another thought experiment. Firstly, consider a transmitter connected via a feeder to a single half-wave dipole. We measure the electric field strength at a particular point in the principal direction and make this our reference field strength E_{ref}. We now reduce the power coming into the dipole to half its original value. The power density is proportional to the square of the electric field strength, so the field strength will reduce to $E_{\mathrm{ref}}/\sqrt{2}$. Now a second dipole is added in line with the first dipole to form a collinear array. The same power is fed to

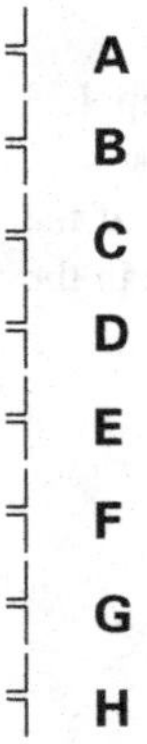

A
B
C
D
E
F
G
H

Figure 2.5 An array of eight, collinear half-wave dipoles.

the second dipole so that the total power from the transmitter is now the same as it was originally but divided between two dipoles. At the point in the principal direction, the two electric fields will add in phase, resulting in an electric field strength equal to $2E_{ref}/\sqrt{2}$, which equals $\sqrt{2}E_{ref}$. The power density will be double the level it was originally when the power was fed to one dipole. The gain of the array is 3 dB greater than the gain of an individual dipole. However, the array is still a passive antenna. The higher gain has been achieved by making the beamwidth narrower. The way this happens is worthy of more detailed investigation.

Figure 2.6 shows an array of two half-wave dipoles. In the principal direction, the path length from each of the dipoles to a distant point is the same. Therefore the electric fields from the two dipoles will add in phase. However, at points away from this principal direction, the path lengths will not be equal. For dipoles separated by a distance d, the path-length difference equals $d \sin \theta$, where θ is the angle between the principal direction and the direction under investigation. If the path-length difference is equal to half a wavelength, then the contributions from the two elements will add in anti-phase and will cancel each other out, producing a null. The first null occurs when

$$d \sin \theta = \lambda/2$$
$$\theta = \sin^{-1}[\lambda/(2d)]. \tag{2.20}$$

If the dipoles are positioned closely together, at a separation of half a wavelength, then the first null will occur where θ equals 90 degrees, as it

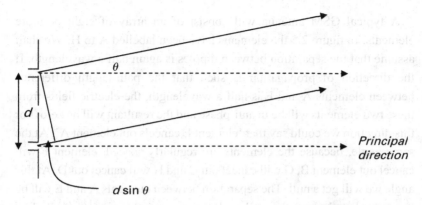

Figure 2.6 An array of two dipoles, showing the path-length difference in directions away from the principal direction.

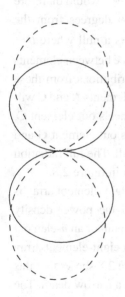

Figure 2.7 Vertical radiation patterns of a half-wave dipole (solid line) and a two-element array (dashed line). The patterns are drawn to the same scale, showing the relative electric field strength assuming equal radiated power. In the principal direction, the electric field strength produced by the two-element array is $\sqrt{2}$ times that of the single dipole.

does with a single dipole, but the fact that the contributions do not add in phase away from the principal direction means that the beam is narrower, resulting in a higher gain.

Figure 2.7 shows a comparison between the radiation pattern of a single half-wave dipole and that of an array of two half-wave dipoles with a separation of half a wavelength.

A typical GSM antenna will consist of an array of eight or more elements. In figure 2.5 the elements have been labelled A to H. We shall assume that the separation between dipoles is again half a wavelength. If the direction of propagation is such that the path-length difference between elements A and E is half a wavelength, the electric fields from these two elements will be in anti-phase and the resultant will be zero. For this direction we could say that 'element E cancels out element A'. At the same angle, because the elements are regularly spaced, element F will cancel out element B, G will cancel out C and H will cancel out D. At this angle we will get a null. The separation between elements A and E will be two wavelengths and the null will occur when $\theta = \sin^{-1}(1/4) = 14°$. There will be another null when the path-length difference equals $3\lambda/2$. At this path-length difference $\theta = \sin^{-1}(3/4) = 49°$. We would therefore expect to see nulls at angles of 14 degrees and 49 degrees from the principal (horizontal) direction. Interestingly, there is a null where $\theta = \sin^{-1}(1/2) = 30°$. Although the path-length difference between elements A and E will be a full wavelength, causing the contributions from these elements to add, the path-length difference between elements A and C will be half a wavelength. So at this angle element A cancels out element C, element B cancels out element D, element E cancels out element G and element F cancels out element H and thus we get a null. The full radiation pattern for an eight-element collinear array is shown in figure 2.8.

In the principal direction, the field strength of the eight-element array is $\sqrt{8}$ times that of the single, half-wave, dipole and the power density produced is therefore eight times that of a dipole. The gain of an n-element array is $(2.1 + 10 \log n)$ dBi. Therefore the gain of an eight-element array would be expected to be approximately 11 dBi. Figure 2.8 shows that this gain is achieved by directing the transmitted power in a narrow beam. The beamwidth between points where the power density reduces to 3 dB below that in the principal direction is 12° in the case of the eight-element array, whereas it is 82° for a half-wave dipole. For a large array with n elements (n greater than 6), the beamwidth is approximately equal to $100/n$ degrees.

When designing a base-station antenna, the beamwidth in the vertical direction cannot be too narrow, or the antenna will not provide sufficient signal strength in the area near the base of its supporting tower.

Figure 2.8 Vertical radiation patterns of a half-wave dipole (dashed line) and an eight-element array (solid line). The patterns are drawn to the same scale showing the relative electric field strength assuming equal radiated power. In the principal direction, the electric field strength produced by the eight-element array is $\sqrt{8}$ times that of the single dipole.

In presenting the radiation patterns for the arrays shown it has been assumed that the signal from the transmitter is fed to the elements such that they are fed in phase with each other. When that is the case the phase difference between the signals received from the various elements at a particular point is due to the difference in path length. It is common, particularly for antennas used for digital mobile radio, for the elements not to be fed in phase. This has the effect of steering the beam in a different direction. It is possible for the phase difference between the elements to be chosen such as to produce a 'down-tilt' (known as electrical down-tilt) in the beam. Figure 2.9 shows the radiation pattern of an eight-element array with electrical down-tilt.

2.7.1 Sectored antennas

It is common for antennas used for digital mobile-radio base stations to use directional, 'sectored' antennas instead of omni-directional antennas. A typical horizontal radiation pattern of a sectored antenna is shown in figure 2.10.

A single base station would provide coverage in all directions by using three sectored antennas pointing at intervals of 120° in the horizontal plane. Figure 2.10 has two lines marked on it to indicate the 120° sector

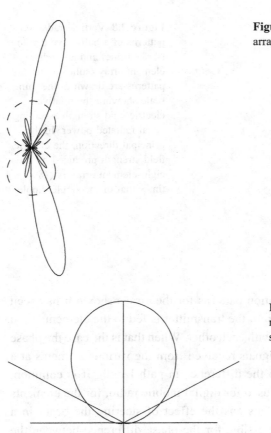

Figure 2.9 An eight-element array with electrical down-tilt.

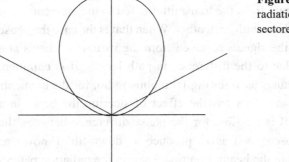

Figure 2.10 The horizontal radiation pattern for a sectored antenna.

that would be served by the particular antenna. The field strength in these directions is much lower than the field strength in the principal direction. The field strength is little more than 30% of its value in the principal direction, which means that the power density is only one tenth (10 dB down) of the principal-direction value. It may seem odd that the antenna is regarded as providing coverage in this direction when the field strength is relatively weak, but the way networks are planned means that the distance over which coverage is required is only half the distance that is required in the principal direction. The Okumura–Hata path-loss models predict a loss that typically depends on distance with a $35.2 \log d$ relationship. That means that, for an increase in distance by a factor of 10, the loss would increase by 35.2 dB. The ratio of distance that would result from a variation in required path loss is given by $10^{\text{loss}/35.2}$, so a decrease in loss by 10 dB would be achieved by reducing the distance coverage by a factor of $10^{10/35.2} = 1.9$. This accommodates well the previous statement that the coverage distance at the angle at which the power density is 10 dB down on the value in the principal direction is only half that in the principal direction. This reduced coverage requirement is a product of the necessity for a mobile radio network to provide continuous coverage. If this is to be achieved then the coverage patterns from the individual base station must fit together to form a continuous pattern (they must tessellate). Figure 2.11 shows how this can be achieved if each sectored antenna has a hexagonal coverage pattern.

The required coverage area for each antenna is a hexagon shape. The length of one side of a hexagon is half the distance across the hexagon from corner to corner. Although it is not possible for a practical antenna to reproduce the sharp corners of the hexagonal coverage area, it is possible to design the antenna to give the best fit possible. Figure 2.12 shows a contour of equal link loss (all points on the contour have an equal link loss to the base station after consideration of antenna gain) for the

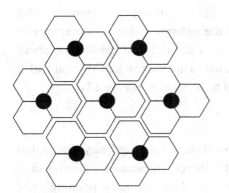

Figure 2.11 Continuous coverage achieved by regularly spaced base stations, each with three sectored antennas. The required coverage area for each antenna is hexagonal in shape.

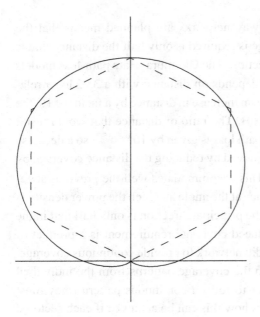

Figure 2.12 The predicted contour of equal link loss using the sectored antenna.

antenna whose radiation pattern is given in figure 2.10 using the Oku-mura–Hata path-loss model. It can be seen that the predicted coverage area is a reasonably good fit to the hexagon.

The gain of a sectored antenna is dependent upon its beamwidth both in the horizontal direction and in the vertical direction. If it consists of eight elements the gain will be approximately 9 dB higher than the gain of the individual element (which will be 2.1 dBi if it is a half-wave dipole). The compression of the energy into a narrow horizontal beam provides further gain. The amount of extra gain depends upon the width of the horizontal beam. If, as in the examples shown, the energy is compressed into about one quarter of the full 360° panorama, about 6 dB of extra gain will be provided. This means that the gain of a sectored antenna will be expected to be approximately 17 dBi.

2.8 Mobile-station antennas

The antennas used in mobile terminals are of a proprietary design, but they are designed to exhibit very little directivity because it is impossible to be sure of the direction in which the mobile antenna is 'pointing'. The lack of directivity translates into a lack of gain and hence it is common to assume that the mobile antenna has a gain of 0 dBi for the purposes of calculating received signal powers.

2.9 Interference and the noise floor

Apart from the basic thermal noise, other forms of noise exist. At frequencies below about 200 MHz, man-made noise (from electronic and industrial sources) and galactic noise add considerably to thermal noise and become the major consideration when determining the required signal level. Above a few hundred megahertz thermal noise dominates and radio planners will use an estimate for this in determining the required signal level on the wanted service. Should the level of man-made noise and interference increase above frequencies of a few hundred megahertz then this will affect the performance of mobile telecommunication systems. This is not unlikely given the increasing speed and number of personal computers and also the increase in the number of portable radio terminals for which no licence is required. If the interference increases the effective noise floor by 1 dB, the range of a base station will decrease by a factor of approximately $10^{1/35} = 1.07$. The area covered by a base station will decrease by a factor of $1.07^2 = 1.14$. Figure 2.13 shows a comparison of coverage areas for a difference of 2 dB in path loss. In this case the coverage range would be reduced by a factor of 1.14 and the coverage area of each base station by a factor of 1.30. The reduction in coverage range would mean that the operator will have to deploy 30% more base stations to provide the same coverage. The situation is even more serious if the rise in the noise floor

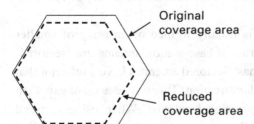

Original coverage area

Reduced coverage area

Figure 2.13 The reduction in coverage area of a sector antenna caused by a 2-dB increase in the noise floor.

occurs after the network infrastructure has been installed. The operator's customers will experience a reduction in the quality of service that will be difficult to rectify. If interference causes the increase in the effective noise floor, installing high-quality receivers will do hardly anything to improve the situation. If the operator has already installed such receivers, they will feel that the investment has been wasted.

2.10 Summary

When considering point-to-area communication it is very common to use the term 'field strength' rather than power density or received power. There is often only a single terminal and antenna (the base station) to consider rather than one at each end of a link. The base station produces a field strength at a particular distance. The received power then depends on the characteristics of the receiving antenna. Predicting the field strength at a distance can be undertaken using a variety of models, most notably ITU-R P.1546. Using this recommendation, it is possible to predict that the field strength produced by a certain transmitted power does not depend greatly on frequency at short distances, but, on longer paths, particularly as the path becomes trans-horizon, the field strength produced reduces with increasing frequency. For point-to-area services, such as digital mobile communications, for which base stations serve a relatively small area, alternative prediction models that generally require less computation are used. One example of this is the Okumura–Hata model. Using this model, it is seen that link loss from the base station to the mobile terminal increases with frequency. One impact of this is that significantly more base stations will be required to provide coverage on a network at 1800 MHz than would be the case if a frequency of 900 MHz were used.

The base-station antenna is usually formed of an array of smaller elements. Two major categories of base-station antenna are 'sectored' and 'omni-directional' antennas. Sectored antennas have a reflector that directs the power in a particular direction. This provides more gain than would an equivalent omni-directional antenna. When using sectored antennas, it is common to find three antennas on a base station providing complete panoramic coverage between them. Mobile-station antennas are omni-directional because requiring the user to point the antenna in a particular direction would be seen as highly inconvenient.

All predictions of coverage are made assuming a certain receiver sensitivity. This in turn assumes a level of background noise. If interference increases the total level of unwanted power at the receiver, this will reduce the coverage from a base station.

3   The effect of obstacles

Predicting the strength of a radio signal in the shadow of an obstacle is a vital function for propagation engineers. The mechanism by which a wave enters into the shadow of an obstacle is known as diffraction. Even the simplest of practical obstacles pose severe mathematical challenges. More easily solved approximations are adopted in order to estimate the strength of diffracted signals. The starting point for diffraction problems is the case where a receiver is in the shadow of a perfectly absorbing 'knife-edge' obstacle. This is then extended to encompass the situation where there are several such obstacles on the path. Many approximate multiple-knife-edge prediction methods exist and the most commonly used are analysed and compared. More accurate 'near-exact' methods are discussed. Although these methods usually make better predictions of the signal strength in the shadow of obstacles, they require significantly more computing time as well as being significantly more complicated to implement. Once an understanding of the properties of a diffracted signal has been obtained, it is possible to derive clearance requirements for a point-to-point path so that diffraction effects may be safely ignored. The insights gained by investigating the mechanism of diffraction into the shadow of an obstacle can be used to analyse two related phenomena: reflection from a finite surface and the formation of the radiation pattern of an aperture antenna.

3.1 Knife-edge diffraction

Diffraction is the name given to the mechanism by which waves enter into the shadow of an obstacle. The phenomenon affects all waves, such as those observed in water, sound waves and radio waves. One theoretical mathematical model that is used to gain greater understanding of the mechanism of diffraction involves the concept of the perfectly absorbing

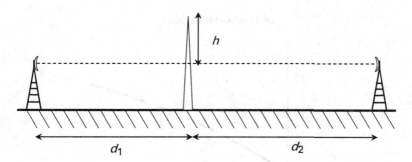

Figure 3.1 The path geometry for knife-edge diffraction

knife edge. The picture used involves a uniform plane wave (such as would come from a very distant source) being incident on a knife edge such that the plane of the edge lines up with the wave front (the plane of the edge is perpendicular to the direction of travel of the wave). It is then possible to predict the strength of the signal in the shadow of the obstacle relative to that without the obstacle. The ratio of the signal strengths without and with the obstacle is referred to as the diffraction loss. The diffraction loss is affected by the path geometry and the frequency of operation. It is possible to absorb all of the relevant factors into a single parameter: the Fresnel parameter. For a simple configuration as shown in figure 3.1, the Fresnel parameter, v, is given by

$$v = h\sqrt{\frac{2}{\lambda}\left(\frac{1}{d_1} + \frac{1}{d_2}\right)}. \tag{3.1}$$

The diffraction loss is then a function of the Fresnel parameter. The method of determining the diffraction loss from the Fresnel parameter is quite complicated, involving the summation of series. However, for values of v greater than about -0.7 the following approximation is valid:

$$\text{loss} = 6.9 + 20 \log\left(\sqrt{(v - 0.1)^2 + 1} + v - 0.1\right) \text{ dB}. \tag{3.2}$$

This function is plotted graphically in figure 3.2.

This reveals some interesting initial findings. If we have 'grazing' incidence ($h = 0$ and hence $v = 0$) then the diffraction loss equals

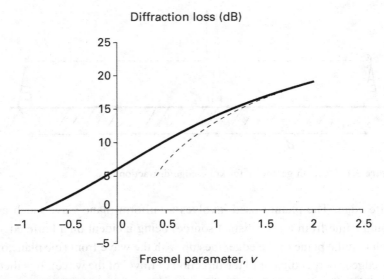

Figure 3.2 Diffraction loss as a function of the Fresnel parameter, v (solid line). The approximation loss $\approx 13 + 20 \log v$ is shown as a dashed line.

approximately 6 dB. That means that the signal strength will fall by 6 dB as the receiver approaches the shadow boundary but before it enters into the shadow region. The Fresnel parameter must be approximately -0.8 (that is, h is negative) before the diffraction loss reduces to zero and the presence of the edge can be ignored.

As the receiving point moves deep into the shadow and v becomes very large, the equation for diffraction loss can be approximated to

$$\text{loss} = 13 + 20 \log v. \tag{3.3}$$

This gives a reasonably accurate answer if $v > 1.5$.

Notice that v is directly proportional to h and inversely proportional to the square root of wavelength. That means that v is directly proportional to the square root of frequency. The higher the frequency, the higher the value of v, all other things remaining equal. Deep in the shadow of an obstacle, the diffraction loss increases with $20 \log v$. That means that loss will increase with $10 \log f$. So, if you double the frequency, deep in the shadow of an obstacle the loss will increase by 3 dB. If you multiply the frequency by 100 (say from 300 MHz to 30 GHz), then the loss will increase by 20 dB.

This establishes a general truth, namely that radio waves of longer wavelength will penetrate more deeply into the shadow of an obstacle.

3.1.1 Huygens' principle and the Cornu spiral

We have thus far studied propagation in free space. The equations we have derived form the foundation of studies into radio wave propagation. However, dealing with transmission from one antenna to another through free space and in the absence of any obstructions does ignore a lot of features that can affect the power received in practice. We are first going to attempt to predict the effect of a single obstacle placed in the path between transmitter and receiver.

In order to do this, we first need a mental picture of the way in which a radio wave travels from the transmitter to the receiver in the absence of any obstruction. To help us here we call upon the theory of wave propagation first put forward in 1690 by Christian Huygens. It remains possibly the greatest insight into the propagation of light (and of electromagnetic waves in general). In order to predict the way in which a wave travels, it is necessary to consider a wave front as the source of an infinite number of 'wavelets'. Effectively, that means regarding each point on a radio wave front as a tiny radio transmitter. The next wave front can then be regarded as the sum of the contributions from all of these tiny transmitters. This means that we can determine the field strength at a receiver by adding together the contributions of all the wavelets at a wave front somewhere between the transmitter and receiver (in principle, any wave front will do). The principle is illustrated in figure 3.3. In the case of an unobstructed path, this just gives us an overcomplicated way of predicting the received power. When there is an obstacle blocking some of these wavelets, then the power of Huygens' principle becomes clear: it gives us a way of predicting the ratio of the received field strengths with and without the obstacle.

It is common to simplify the situation by considering a two-dimensional picture. Further, the path is considered to be 'long' so that the wave front can be considered to be a straight line (rather than part of a circle). This wave front is divided into wavelets.

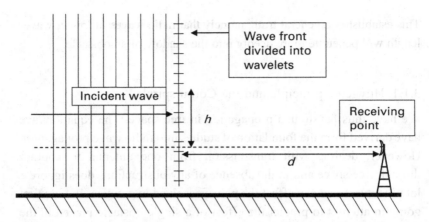

Figure 3.3 An illustration of Huygens' principle. The received signal is the phasor sum of the contributions from all the wavelets on a particular wave front.

The process is continued by trying to imagine the effect of each wavelet at a particular receiving point. The distance from each wavelet to the receiving point is different and therefore the contributions of each wavelet would not arrive at the receiving point in phase. The way in which they add can best be visualised by a phasor diagram. (The resultant of two sine waves can be determined in a similar way to the resultant of two vectors, such as forces. The phasor diagram assists in doing this in a similar manner to a vector diagram. More information on phasor addition can be found in Appendix 2). The position of any wavelet along the wave front is indicated by the distance from the point where the perpendicular from the receiving point meets the wave front. Let this distance be h. As h increases, the distance to the receiving point increases. If the length of the perpendicular from the receiving point to the wave front is d then the distance from any wavelet is given by $\sqrt{h^2 + d^2}$. At this point an approximation is introduced, which is valid only if $d \gg h$. Another way of regarding this is that the angle between the perpendicular and the line joining the wavelet to the receiving point must be small:

$$\sqrt{d^2 + h^2} = d\sqrt{1 + h^2/d^2}$$

$$\approx d + \frac{h^2}{2d}. \tag{3.4}$$

Thus the distance to the receiving point increases with the square of the value of h. When adding the contributions of the different wavelets we need to perform what is known as 'phasor' addition. This simply means that we have to consider any phase difference between the contributions from the wavelets. The process is similar in practice to vector addition: two equal phasors with 180° phase difference will add to zero in the same way as two equal vectors, such as forces, acting in opposite directions will add to zero. For further information on phasor addition, refer to the appendix. The phase difference between the contributions of any two wavelets is proportional to the difference in path lengths between them and the receiving point. A path-length difference of a full wavelength produces a phase difference of 360° or 2π radians. In general, the phase difference between the contributions of any two wavelets at positions h_1 and h_2 above the perpendicular from the receiver to the wave front is given by

$$\text{phase difference} = \frac{\pi(h_2^2 - h_1^2)}{\lambda d} \text{ radians}, \tag{3.5}$$

where λ is the wavelength.

The relative phase of the contribution is proportional to the square of h, d being a constant factor in the distance to all wavelets. The wavelet needs to have a 'small but finite' width. 'Small but finite' is a commonly heard phrase in radio propagation analysis. In this case, the arithmetic is simpler if we consider the variation in phase across the surface of a wavelet to be negligible (this triggers a debate regarding the meaning of 'negligible'; in practice you make the phase variation smaller and smaller until the effect on the resulting prediction is so small that any further effort is deemed unproductive). We are now going to build up a phasor diagram that shows the contribution from wavelets that have equal phase variation (you can think of it as going up in perhaps 5-degree steps). The wavelets will not all have the same width since the phase shift is proportional to $h_2^2 - h_1^2$ whereas the width of the wavelet equals $h_2 - h_1$. If we want to keep $h_2^2 - h_1^2$ constant, then $h_2 - h_1$ will get smaller as h increases. This means that the contribution of each wavelet will reduce with increasing phase shift. The effect of this is to produce a

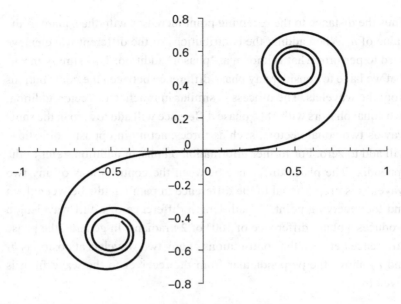

Figure 3.4 A phasor diagram summing wavelet contributions.

phasor diagram that spirals towards a fixed point. If the wavelet contributions were equal, the phasor diagram would just produce a circle.
The spiral produced is shown in figure 3.4.

The spiral shown is known as the Cornu spiral. The x-axis and the
y-axis are simply indicators of the relative phase angles of the contributing wavelets. They have a scale but one of the beauties of this spiral is
that the units are not relevant. We use the diagram to predict the strength
of the signal in the shadow of an obstacle relative to that when no
obstacle exists. On examining the spiral, we see that it spirals in at one
end towards the point $(-0.5, -0.5)$ and, at the other, towards the point
$(0.5, 0.5)$. Thus, the phasor that represents the strength of the signal
without any obstacle extends from the point $(-0.5, -0.5)$ to the point
$(0.5, 0.5)$. The length of this phasor on the phasor diagram is $\sqrt{2}$. The
relative strength of any signal in the presence of an obstacle is given by
the length of the line from the point on the spiral representing the
location of the edge of the obstacle divided by $\sqrt{2}$. The last trick that we
have to perform is to identify the point on the spiral that corresponds to

the location of the edge of the obstacle. It is fairly obvious that this corresponds to the phase difference between the wavelet at the edge and the perpendicular from the receiving point to the wave front (the '$h = 0$' point). The phase difference at the edge of an obstacle located at a general distance from the perpendicular is given by $\pi(h^2)/(\lambda d)$. Suppose that for a particular geometry the relative phase difference was $\pi/2$ radians (90 degrees). This would be identified with a particular point on the Cornu spiral. The strength of the diffracted signal compared with the free-space signal can be determined using the ratio of the lengths of the two phasors shown in figure 3.5. The lengths of the phasors represent electric fields rather than powers. We have seen from equation (3.3) that the power density is proportional to the square of the electric field strength. Because of the need to square the field strength, conversion to dB leads to the difference following a '20 log' relationship rather than a '10 log' relationship where the ratios are powers. More detailed information on decibels (dB) is given in the appendix.

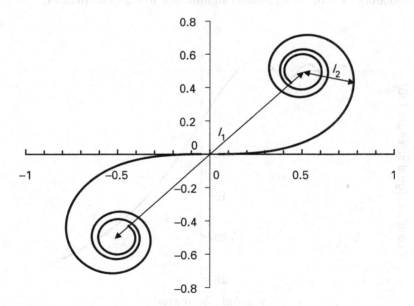

Figure 3.5 A Cornu spiral indicating how diffraction loss may be determined when the phase shift from the top of an obstacle in the figure is 90 degrees relative to the direct line.

Again referring to figure 3.5:

$$\text{diffraction loss} = 20 \log\left(\frac{l_1}{l_2}\right) \text{ dB.} \tag{3.6}$$

In this case the ratio of the two lengths is approximately $5:1$, suggesting that the diffraction loss would be approximately 14 dB.

An examination of the spiral allows us to make some interesting deductions. For example, if the knife edge were exactly at the perpendicular joining the wave front to the receiving point, the length, l_2, would be exactly half that of l_1, leading to a predicted diffraction loss of 6 dB. Note also that the distance from some points on the spiral to the $(0.5, 0.5)$ point is greater than $\sqrt{2}$. For these circumstances a signal stronger than that under free-space conditions will be received. Figure 3.6 shows the diffraction loss as a function of the Fresnel parameter that extends the values of v below -0.80, showing the fact that the path loss becomes negative. Although it is possible to demonstrate this phenomenon in the laboratory, it is of little practical significance to the radio planner.

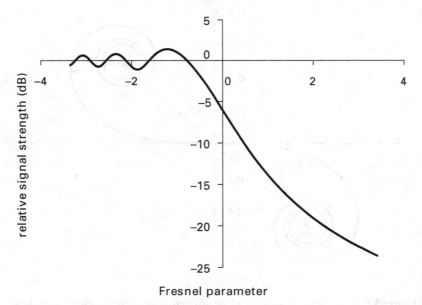

Figure 3.6 The signal strength against the Fresnel parameter, showing that the diffraction loss can become negative.

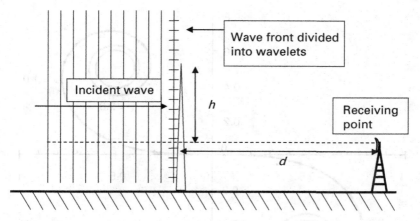

Figure 3.7 The geometry of a plane wave incident on a knife-edge obstacle.

The method shown here underpins methods of predicting the strength of the diffracted field. However, determining the relative phase difference for a particular geometry and then making measurements on the Cornu spiral can be time-consuming. A faster method involves using the Fresnel parameter. Where a plane wave is incident upon the diffracting obstacle, as shown in figure 3.7, the Fresnel parameter, v, is given by

$$v = h\sqrt{\left(\frac{2}{d\lambda}\right)}.$$ (3.7)

To compute the diffraction loss it is, strictly speaking, necessary to compute the 'Fresnel integral', but the following approximation is very useful:

$$\text{loss} = 6.9 + 20\log\left(\sqrt{(v-0.1)^2+1} + v - 0.1\right) \text{dB}.$$ (3.8)

This equation is valid for values of v greater than about -0.7.

For large values of v (greater than about 1.5), the diffraction loss is approximately equal to $13 + 20\log v$ dB.

It is possible to mark the value of v along the Cornu spiral. Since equal increments of v correspond to equal increments of h, they will represent equal increments along the spiral. Figure 3.8 shows the Cornu spiral with markers indicating the value of v in increments of 0.2.

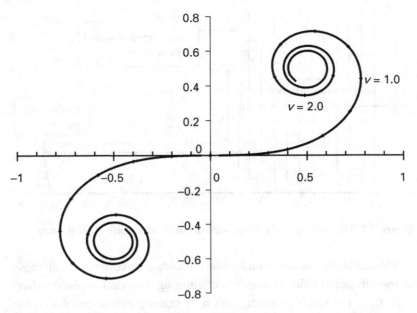

Figure 3.8 A Cornu spiral showing values of the Fresnel parameter, v, at increments of 0.2.

Note that, when $v = 1.0$, the spiral indicates a phase shift of 90 degrees compared with $v = 0$ and, when $v = 2.0$, the phase shift is 360 degrees. This indicates the general rule that the

$$\text{phase shift} = 90 \times v^2 \text{ degrees.} \tag{3.9}$$

Further note that, for small values of v, the distance from the origin is approximately equal to the value of v.

The assumption of the wave incident on the diffracting edge being a plane wave is often not valid. For a more general geometry as shown in figure 3.1

$$v = h\sqrt{\frac{2}{\lambda}\left(\frac{1}{d_1} + \frac{1}{d_2}\right)}. \tag{3.10}$$

Notice that, if one of the distances is taken to be infinity, the equation reduces to the simpler form

$$v = h\sqrt{\left(\frac{2}{d\lambda}\right)}. \tag{3.11}$$

Note that, to use these equations, you must use 'self-consistent' units. All parameters on the right-hand side of the equation are lengths or distances. They must all have the same units (usually metres).

Example: Considering the geometry of figure 3.1, we shall determine the diffraction loss incurred for $d_1 = 10$ km, $d_2 = 5$ km and $h = 20$ metres. We shall perform the calculation at frequencies of 1 GHz and 10 GHz. At 1 GHz, $\lambda = 0.30$ metres and therefore

$$v = 20\sqrt{\frac{2}{0.3}\left(\frac{1}{10000} + \frac{1}{5000}\right)} = 0.89$$

and the diffraction loss will be given by

$$\text{loss} = 6.9 + 20\log\left(\sqrt{(0.89 - 0.1)^2 + 1} + 0.89 - 0.1\right) = 13.2 \text{ dB}.$$

At 10 GHz, the wavelength equals 0.03 metres, the Fresnel parameter

$$v = 20\sqrt{\frac{2}{0.03}\left(\frac{1}{10000} + \frac{1}{5000}\right)} = 2.83$$

and the diffraction loss equals

$$6.9 + 20\log\left(\sqrt{(2.83 - 0.1)^2 + 1} + 2.83 - 0.1\right) = 21.9 \text{ dB}.$$

This demonstrates the general trend that, as frequency increases, so does diffraction loss. Lower-frequency signals will penetrate more into the shadow of obstacles.

3.2 Clearance requirements

Note that, for the situation where the line joining the transmitter and receiver just touches the peak of the obstacle, the parameter v is zero. At this point the diffraction loss is 6 dB at all frequencies. When designing a point-to-point link, some clearance is necessary so as to ensure that diffraction losses are negligible. The diffraction loss does not reduce to zero suddenly as line of sight is achieved between the transmitter and the receiver. Rather, a certain amount of clearance is required. The loss reduces to zero when v is approximately -0.8. Arranging for v to be no greater than -0.8 (or perhaps -1.0) is seen as an appropriate recommendation.

3.3 A résumé of the assumptions and approximations

One assumption made is that the knife edge has no thickness and yet is perfectly absorbing – a physical impossibility. Further, in the space immediately adjacent to the knife edge on the side of the receiver, the electric field is 'full strength' just above the knife edge and zero just below it. This implies an infinite electric field gradient: another physical impossibility. Additionally, the diffraction angle must be small for the equations regarding phase difference to be valid. Making assumptions that turn out to be physical impossibilities may sound very unsafe, but the reality is that a detailed investigation into diffraction by a sheet of a good electrical conductor reveals that the differences are small. The solution for diffraction by a perfectly absorbing knife edge remains a cornerstone of diffraction studies.

3.4 Diffraction through an aperture

Thus far we have considered diffraction over a single knife edge. We have had to integrate from the edge of the obstacle to infinity. It is possible to use the Cornu spiral to predict the received signal strength when an electromagnetic wave passes through an aperture, or window, formed of two parallel knife edges.

To use the spiral you calculate the point on the spiral for each knife edge as if it were the only knife edge. Then the line joining these two points together is a phasor representing the field strength received on the other side of the aperture. As an example consider the geometry of figure 3.9.

Suppose that the frequency of transmission is 1 GHz, giving a wavelength of 0.3 metres. Let d equal 550 metres, h_1 equal 9 metres and h_2 equal 18 metres. The path-length difference between the lower edge and the perpendicular is approximately 0.075 metres, or a quarter of a

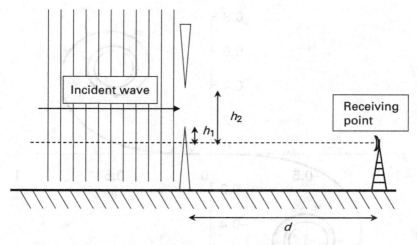

Figure 3.9 The geometry for diffraction by an aperture.

wavelength, giving a phase difference of 90 degrees. Since h_2 equals twice h_1 the phase difference at the upper edge is 360 degrees. These points can be marked on the Cornu spiral and a line representing the electric field strength of the wave passing through the aperture drawn, as shown in figure 3.10.

The distance between the two points is measured as about one fifth of the free-space level, suggesting that the power received at the receiving point would be about 14 dB less than that if the path were unobstructed. Examining figure 3.10 shows that the first edge corresponds to a value for the Fresnel parameter, v, of 1 and the second edge corresponds to v equals 2 (refer to figure 3.7 for confirmation of values of v at points on the spiral). Still referring to figure 3.10, suppose that the size of the aperture were to be increased by raising the upper obstruction such that its edge corresponded to a value for v of 2.2. Oddly, now the line joining the point $v = 2.2$ on the spiral to $v = 1$ is only about half of that joining $v = 2.0$ to $v = 1$. Thus, if the size of the aperture were increased by a small amount, the received signal would decrease by about 6 dB. This is an odd prediction but it is verified by laboratory experiments. If the aperture were further increased, to a point where $v = 2.6$, the received signal would increase to a maximum before reducing again should the aperture be

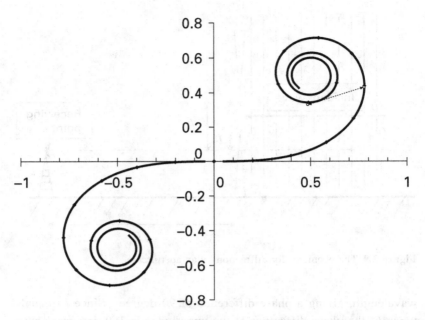

Figure 3.10 A Cornu spiral, with a dotted arrow indicating the phasor magnitude if an aperture allows wavelets between phases of 90 and 360 degrees to pass through.

made larger again. As the upper obstacle is raised the received signal strength will oscillate, gradually settling at the value it would be if only the lower diffracting edge were present.

Diffraction through an aperture may seem to be a rather rare event, but appreciation of this can lead to a deeper understanding of two more common phenomena: reflection from a surface of finite size and the nature of radiation from an aperture antenna.

3.5 Reflection from a finite surface

The method used to predict the reflected field is to consider a virtual image of the transmitter. If the reflecting surface is not infinite (or does not occupy all of the area between the transmitter and receiver) then the reflected field cannot be predicted by simply knowing the magnitude of the reflection coefficient at the surface. The equivalence of reflection from a finite aperture is illustrated in figure 3.11. The 'view' from the receiver of

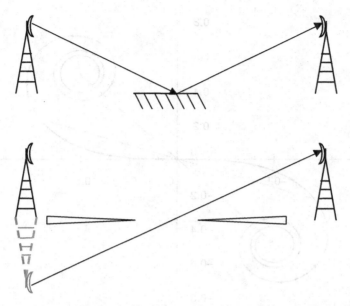

Figure 3.11 The equivalence of reflection from a flat surface and diffraction through an aperture.

the reflection of the transmitter in the reflecting surface is the same as its view through the aperture of the virtual image of the transmitter.

One question of interest is: 'How big must the reflecting surface be to have a similar effect to an infinite area?'. Again the Cornu spiral can help us. It can be seen that, if the surface is large enough to include phase shifts of ±60 degrees on either side of the reflecting point (see figure 3.12) then the received signal will be approximately the same as if the reflecting surface were of infinite size. This corresponds to an excess path length of one sixth of a wavelength. Suppose that, in figure 3.10, the path length is 20 km and the height of the antennas above the reflecting surface is 30 metres. At a frequency of 10 GHz, the wavelength will be 3 cm and the path-length difference between the direct line and one going via the edge of the reflecting surface need be only 5 mm.

This may seem a very small amount, but the geometry of the situation (with very small angles between the direction of propagation and the plane of the reflecting surface) means that the size of the reflecting

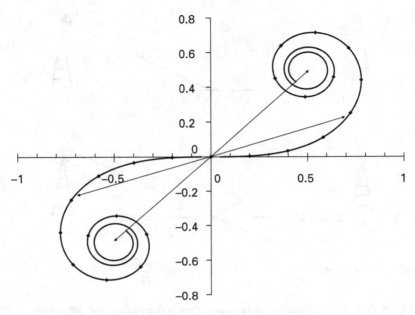

Figure 3.12 A Cornu spiral showing that the signal strength beyond an aperture with a phase shift of ±60° will be as strong as the free-space level.

surface may be surprisingly large. Let the width of the reflecting surface be $2x$. The length of the direct path is

$$\sqrt{20\,000^2 + 60^2} = 20\,000.09 \text{ metres.}$$

The length of the ray via the edge of the reflecting surface is

$$\sqrt{(10\,000 + x)^2 + 30^2} + \sqrt{(10\,000 - x)^2 + 30^2}.$$

The difference between the two distances equals 0.005 metres when x equals 2300 metres. This means that the reflecting surface must be 4600 metres wide to give a reflected signal as big as if the surface occupied the whole distance between the transmitter and receiver. At first, this result is quite surprising. A surface of length 4.6 km produces a path-length difference of only 5 mm from the extreme of the surface to the centre. Suppose that the reflecting surface was only 1 km in length. In that case, the dimension x would be 500 metres and the path-length difference would be only 0.22 mm for a ray travelling via the edge of the surface compared with one travelling via the point of reflection. This corresponds to 0.0075 wavelengths. The Fresnel parameter $v = \sqrt{4 \times 0.0075} = 0.17$. Thus the strength of the reflected field can be determined by comparing the length of the phasor connecting $v = \pm 0.17$ on the Cornu spiral with the length from one end of the spiral to the other (a length of $\sqrt{2}$). For a small value of v the distance from the origin is almost equal to the value. Thus the length of the phasor connecting will be approximately 0.34. The reflected signal at the receiver, even if the surface is a perfect reflector, will be $20 \log(\sqrt{2}/0.34) = 12.2$ dB lower than the direct signal. If the reflection coefficient has a magnitude of less than unity, the difference between the amplitudes of the direct and reflected signals will be larger.

3.6 Antenna radiation patterns

Information in antenna data sheets usually contains two radiation patterns: vertical and horizontal. For a microwave 'dish' antenna the vertical and horizontal radiation patterns are usually nearly identical: the radiation pattern has circular symmetry.

The radiation pattern does not consist of a single 'beam'. There will be additional smaller beams known as side lobes. These are particularly noticeable with antennas that transmit with a very narrow beam, such as those used for microwave point-to-point links. These side lobes are generally problematic in radio communications because they represent the fact that not all of the energy travels in the direction intended (and also, that antennas can receive signals from directions other than that in which they are pointed). This is a highly significant factor contributing to the level of interference in a radio network. The ratio of the gain in the main beam (what we have simply been calling the antenna gain) and the gain in the most significant side lobe is a significant antenna parameter in allowing the planner to assess how much interference will be caused in a network.

3.6.1 The radiation pattern of an aperture antenna

In chapter 2 we examined the vertical radiation pattern of an omni-directional antenna. This pattern was predicted by determining the phase difference between contributions from the various half-wave dipoles that made up the antenna. Although there are no elements as such with an aperture antenna, Huygens' principle allows us to adopt a similar approach when considering the radiation pattern of these antennas. We will attempt to predict the 'far-field' radiation pattern. This is the radiation pattern that would be measured at a great distance from the antenna (in theory, at an infinite distance). One assumption that is a great help in predicting this radiation pattern is that of the equivalence between a parabolic dish illuminated by a source at its focus and an aperture illuminated from behind by a uniform plane wave. This equivalence is illustrated in figure 3.13. The important feature of the paraboloid shape of the reflecting dish is that it equalises the path lengths from the focus to points on a plane in front of the dish perpendicular to the principal direction. Because the path lengths are equal, the wave travelling forwards immediately in front of the dish is similar to a uniform plane wave. This is the same as the wave immediately in front of the aperture that is illuminated from behind by a 'genuine' uniform plane wave. In both cases the wave will disperse as it progresses.

If we consider a side view of the aperture illuminated from behind, it gives an idea of the way in which side lobes will be produced. If we

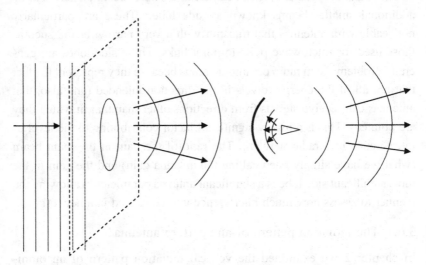

Figure 3.13 The equivalence of an aperture illuminated from behind by a plane wave and radiation from a parabolic dish antenna.

divide the plane wave into wavelets in the plane of the aperture, we can approach the problem in a similar manner to that used to predict the radiation pattern of an array of half-wave dipoles. In the principal direction, all the contributions from the wavelets will add in phase. However, there will be a null in a direction such that the variation in path length over the face of the aperture equals one wavelength. In this direction, each wavelet will have a corresponding wavelet for which the path-length difference is half a wavelength and they will therefore cancel out. This will occur when the angle is such that

$$\sin \theta = \lambda/d. \tag{3.12}$$

As the angle increases we come out of the null. There is a second null when

$$\sin \theta = 2\lambda/d \tag{3.13}$$

(in general there is a null where $\sin \theta = n\lambda/d$, where n is any integer).

Between these two nulls, there is a peak in signal strength, the first side lobe. In the direction of this peak, the contributions from each wavelet do not add in phase and therefore the peak will not be as strong as that in the principal direction, but neither do the contributions cancel out. Figure 3.14 shows the radiation pattern for an evenly illuminated aperture where the aperture width is approximately four wavelengths.

Two side lobes on either side of the main beam are clearly visible. The peak of the first side lobe shows that it would produce a field strength of approximately 21% of that produced in the principal direction. That corresponds to a power density 14 dB less than that in the principal direction. The amount of energy in the side lobes can be reduced by tapering the illumination of the aperture towards its edge. In a practical dish antenna, the feeder at the focus will illuminate the aperture and designing the optimum illumination of the dish is a necessary activity if

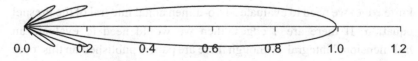

Figure 3.14 The radiation pattern of an evenly illuminated aperture antenna with diameter approximately equal to four wavelengths.

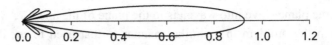

Figure 3.15 The radiation pattern of an aperture antenna for which the aperture illumination is tapered towards its edges.

the antenna itself is to have good performance. Figure 3.15 shows the radiation pattern for a similar antenna to that of figure 3.14 but with an aperture illuminated such that the contributions from wavelets near the edge of the aperture will be less than those from wavelets near the centre.

Notice that the side lobes are much smaller than when the aperture illumination is tapered. The two radiation patterns are drawn to the same scale and reveal that the gain in the principal direction has been reduced (by about 0.8 dB) by tapering the illumination of the aperture.

3.7 Multiple diffracting edges

When determining appropriate equations to use for calculating the diffraction loss caused by a single knife-edge obstacle, several 'not-quite-true' assumptions had to be made. For example, the region above the knife edge is assumed to be perfectly evenly illuminated by the incident wave (the incident wave is 'uniform') and the diffraction angle is assumed to be 'small'. These assumptions are not so ridiculous as to make the equation invalid. However, when a second knife edge is placed in the shadow of the original knife edge and the receiving point is in the shadow of this second knife edge, the situation becomes much more complicated. The signal strength in the shadow of the first obstacle increases with height and therefore the wave illuminating the second knife edge will not be uniform. This means that any additional loss caused by the second knife edge will be extremely complicated to predict. Indeed, to do the job properly (or as properly as we did for the single knife edge) we need to evaluate a two-dimensional integral of the Fresnel equation. If there are n edges, then we would need to evaluate an n-dimensional integral. Although there are papers published on this topic putting forward methods of determining the value of such an integral, they are complex to implement and demanding on computer resources. It is common for approximate methods to be used. The three most commonly used approximate methods are the Bullington [3] method, the Deygout [4] method and the Epstein–Petersen [5] method. None of them is perfect. Errors introduced with the Epstein–Petersen and Deygout methods tend to increase as the number of diffracting edges becomes larger. Most notably, both of these methods over-predict the diffraction loss when there is a large number of edges that are almost aligned with each other (such as streets of terraced houses with pitched roofs). The Bullington method actually reduces all knife edges to a single 'equivalent' knife edge. This has the problem of ignoring obstructions that will, in practice, affect the strength of the received signal. When observing a practical terrain and processing this for use by one of the methods, the radio planner has particular problems. Real terrain profiles do not have the appearance of well-defined knife-edge obstacles. Rather, there may be a general roughness to an obstructing obstacle. Even determining the number of diffracting edges to consider may be a challenge. It is common for a maximum number of edges to be considered (often this maximum number is as small as three) when estimating the strength of the diffracted field.

3.7.1 Comparison of methods

Determining the strength of a signal in the shadow of a single knife edge necessitates solving the Fresnel integral, or using an approximation to it. Where there are multiple diffracting edges, it is necessary to solve a multi-dimensional Fresnel integral. Methods for doing this have been put forward but are extremely computationally intensive and it is almost universal for some form of approximation to be used. The uniform theory of diffraction (UTD) can be adapted for multiple diffractions and there are three further commonly used methods that are known after their developers and are called the 'Bullington', 'Epstein–Petersen' and 'Deygout' methods, respectively.

The Bullington method
A complex terrain profile is reduced to a single knife edge when the Bullington method is implemented. The location of the knife edge is the

point at which the extended lines joining the transmitter and receiver to their respective dominant obstacles (the obstacle that is at the greatest angle of elevation as viewed from transmitter or receiver) meet.

The Epstein–Petersen method

The link is divided into several 'hops', each involving just one diffracting edge. The total diffraction loss is computed by adding the decibel loss of each diffracting edge on its particular 'hop'. Note that the hops overlap each other.

The Deygout method

Although similar in nature to the Epstein–Petersen method, it will produce different results in most circumstances. The procedure starts by determining the dominant edge, that is the edge that would produce the most diffraction loss if it were the only obstruction. The diffraction loss that this edge would produce is determined. Then the top of this edge becomes a virtual transceiver dividing the link into two. On each of these sections, the dominant edge (if there is one) is again determined and the process is repeated. The total loss is the decibel sum of the losses on each of the sections identified.

Example calculations

As an example, we shall estimate the diffraction loss on the link whose geometry is shown in figure 3.16 using the Epstein–Petersen, Deygout and Bullington methods. The frequency is assumed to be 600 MHz ($\lambda = 0.5$ m).

First, using the Epstein–Petersen method we will evaluate the hop between the left-hand antenna and the middle knife edge. This hop is obstructed by the left-hand knife edge that is 30 metres above the horizontal line joining the two antennas. At the point of the obstruction the line joining the left-hand antenna to the middle knife edge will be $50 \times 7/12 = 29.17$ metres above the horizontal line. Thus the obstruction will be only 0.83 metres above the line joining the antenna to the middle knife-edge.

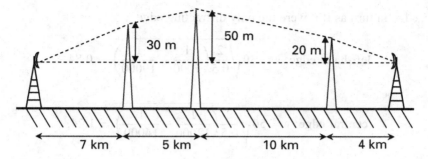

Figure 3.16 A radio path obstructed by three knife-edge obstacles.

Now

$$v = h\sqrt{\frac{2}{\lambda}\left(\frac{1}{d_1}+\frac{1}{d_2}\right)} = 0.83\sqrt{\frac{2}{0.5}\left(\frac{1}{7000}+\frac{1}{5000}\right)} = 0.031.$$

Using the equation

$$\text{loss} = 6.9 + 20\log\left(\sqrt{(v-0.1)^2+1} + v - 0.1\right)$$

gives a loss for this hop of 6.3 dB.

Now consider the second hop from the left-hand knife edge to the right-hand knife edge, a path of 15 km that is obstructed by the central knife edge. The central knife edge will protrude a distance of $20 + 10 \times 5/15 = 23.3$ metres, giving a value for the parameter of 0.81 and a diffraction loss of 12.6 dB.

The final hop is from the central knife edge to the right-hand antenna. This path is obstructed by the right-hand knife edge. This knife edge will protrude into the line of sight by a distance of $20 - 50 \times 4/14 = 5.7$ metres, giving a value for the parameter v of 0.21 and a diffraction loss of 7.9 dB. The total loss would be estimated as the sum: $6.3 + 12.6 + 7.9 = 26.8$ dB. Remember that this would be added to a prediction for the free-space loss, which for a distance of 26 km and a frequency of 600 MHz would be 116.3 dB, giving a total path loss of 143.1 dB.

We shall now re-estimate the diffraction loss using the Deygout method. The first step is to compute the Fresnel parameter, v, for each edge in turn as if it were the only diffracting edge:

$$\text{left-hand edge: } v = 30\sqrt{\frac{2}{0.5}\left(\frac{1}{7000}+\frac{1}{19000}\right)} = 0.84,$$

$$\text{central edge: } v = 50\sqrt{\frac{2}{0.5}\left(\frac{1}{12000}+\frac{1}{14000}\right)} = 1.24,$$

$$\text{right-hand edge: } v = 20\sqrt{\frac{2}{0.5}\left(\frac{1}{22000}+\frac{1}{4000}\right)} = 0.69.$$

Thus the central edge is the 'dominant edge'. The diffraction loss if this were the only edge would be

$$\text{loss} = 6.9 + 20\log\left(\sqrt{(1.24 - 0.1)^2+1} + 1.24 - 0.1\right) = 15.4\,\text{dB}.$$

The next procedure is to calculate any additional loss caused by obstacles between this dominant edge and the left-hand and right-hand antennas, respectively. These losses have already been computed as part of our calculations for the Epstein–Petersen method. The diffraction loss on the path from the left-hand edge to the central edge was determined to be 6.3 dB and the loss from the central edge to the right-hand edge was determined to be 7.9 dB. Thus the total loss = 15.4 + 6.3 + 7.9 = 29.6 dB. This must be compared with a path loss of 26.8 dB predicted using the Epstein–Petersen method. So, which is right? They are each approximations of an exact solution of the two-dimensional diffraction path but neither will be exact. The Epstein–Petersen methodology appears, perhaps, more intuitively correct for the path geometry considered here. However, if the geometry is different, the Epstein–Petersen method poses some awkward questions. In figure 3.17, the left-hand diffracting edge is now not as high as in the previous situation. In fact, it no longer obstructs the line of sight between the left-hand antenna and the central edge, but could still produce a diffraction loss on this path. The Deygout method would simply result in a lower diffraction loss being computed.

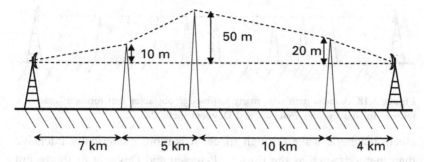

Figure 3.17 A radio path obstructed by three knife-edge obstacles but with a line of sight from one antenna to the central obstacle.

The Epstein–Petersen method has the dilemma of requiring the peak of the left-hand edge to become a virtual source for the next hop. This would result in the predicted diffraction loss due to the central edge being significantly larger. In this case, the Deygout method is probably the better choice. In fact, the Deygout method is probably chosen more often than the Epstein–Petersen method.

In the geometry shown in figure 3.16, both the Deygout method and the Epstein–Petersen method can be expected to over-estimate the diffraction loss. This is because the field strength at the apex of the second obstacle is used to predict the diffracted field in the shadow of that obstacle, although the equation used is valid only if the incident field is uniform (i.e. does not vary with height). The field strength in the shadow of the first obstacle will increase with height. A similar argument can be applied to the diffracted field in the shadow of the second obstacle. It is tempting to ask what the 'exact' solution to this problem would be. But there is no genuinely exact mathematical solution. The best one could think of is to use the transmitter and the obstacles to define boundary conditions and solve Maxwell's equations at each point. This method inevitably involves dividing the space into a finite grid with the grid size affecting the result. Further, it is very resource-intensive and even modern computers would struggle to solve such problems within a realistic timescale. One near-exact method that has been used successfully is known as the 'parabolic-equation' method and is used to 'step' an

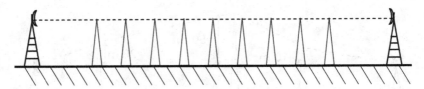

Figure 3.18 A radio path with many knife-edge obstacles that forms a 'case of failure' for the Epstein–Petersen and Deygout methods.

electromagnetic wave through space. It is more numerically intensive than methods such as the Epstein–Petersen and Deygout methods, but most general diffraction problems are tractable. It has been developed by Craig [6]. Professor Craig has been kind enough to analyse the problem of figure 3.16 and reports a predicted diffraction loss of 22.4 dB: significantly lower than that predicted either by the Epstein–Petersen or by the Deygout method. One advantage of the parabolic-equation method is that it can be used to predict the effect of atmospheric structure by virtue of its being able to consider varying refractive index. It reveals a fascinating picture of the way in which an electromagnetic wave will propagate when the atmosphere assumes unusual structures.

However, both the Epstein–Petersen method and the Deygout method have some interesting cases of failure. Figure 3.18 shows many knife-edge obstacles whose apexes just touch the line joining the two antennas. The loss for a single knife edge when the line joining the transmitting and receiving antennas just touches the apex is 6 dB. Both the Epstein–Petersen method and the Deygout method would predict a loss of $6n$ dB for n such knife edges. In the example shown a diffraction loss of 54 dB would be predicted, whereas a more accurate estimation would be approximately 15 dB.

Example prediction using the Bullington method

Figure 3.19 shows how the single equivalent knife edge is constructed in the Bullington method. Only two knife edges are ever considered in the determination of the equivalent single knife edge. In the case considered here it can be seen that the central knife edge is ignored, whereas in the Deygout method it was identified as the most significant edge.

We use geometry to find the height of the equivalent single knife edge above the line joining the two antennas and its distance from one end. Let

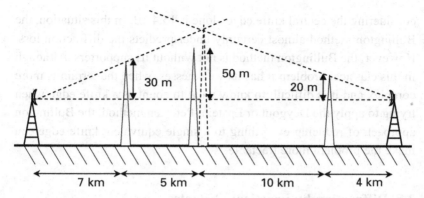

Figure 3.19 An illustration of the Bullington method as applied to a three-edge diffraction problem.

x be the distance in kilometres from the left-hand antenna:

$$\frac{30}{7}x = \frac{20}{4}(26 - x),$$

$$6x = 7(26 - x),$$

$$6x = 182 - 7x,$$

$$x = 182/13 = 14.$$

The height, h, can then be determined:

$$h = 30\frac{182}{7 \times 13} = 60 \text{ metres.}$$

The Fresnel parameter, v, can now be determined.

$$v = h\sqrt{\frac{2}{\lambda}\left(\frac{1}{d_1} + \frac{1}{d_2}\right)} = 60\sqrt{\frac{2}{0.5}\left(\frac{1}{14\,000} + \frac{1}{12\,000}\right)} = 1.49.$$

Now we can determine the diffraction loss from

$$\text{loss} = 6.9 + 20\log\left(\sqrt{(1.49 - 0.1)^2 + 1} + 1.49 - 0.1\right) = 16.7 \text{ dB.}$$

Thus, the Bullington method predicts a diffraction loss significantly lower than that predicted by the other two methods (and by the parabolic-equation method). Indeed, the single equivalent knife edge is not much higher than the actual central knife edge. The loss predicted by considering the central knife edge alone is 15.4 dB. In this situation, the Bullington method almost certainly under-predicts the diffraction loss. However, the Bullington method is not without its supporters. Although in this classical problem it has its weaknesses, when the terrain is more complex and it is difficult to know what to count as a knife edge when trying to apply the Deygout or Epstein–Petersen method, the Bullington approach of reducing everything to a single equivalent knife edge can make the problem more tractable.

3.8 Diffraction by practical obstacles

The analysis has so far been restricted to the case where the diffracting obstacles are sharp knife edges. Experiments to verify these analyses have been done under laboratory conditions with thin sheets of aluminium or copper (or even razor blades) used as obstacles. In practice radio engineers need to be able to predict the signal strength in the shadow of real hills. These will not be true knife edges and often have a finite thickness. This means that instead of just diffraction the process will involve diffraction at one side of the obstacle, propagation along the top of the obstacle and diffraction at the other side of the obstacle. This will inevitably increase the loss at each obstacle. The signal strength in such circumstances can be predicted only with a limited accuracy. The overestimation of diffraction loss by the Deygout and Epstein–Petersen methods when the obstacle is a perfect knife edge can in practice compensate for the fact that real obstacles are not perfect knife edges.

3.9 The GTD/UTD

The geometric theory of diffraction (GTD) provides a method of predicting the signal strength in complex physical situations. It does this by introducing the concept of a 'diffracted ray', hence the GTD is part of the family of 'ray-tracing' methods of predicting the signal strength. The first part of the prediction method involves identifying all major 'propagation mechanisms' whereby the radio wave can travel from the transmitter to the receiver (such as by diffraction, reflection or a combination of the two) and then evaluating the contribution from each mechanism in isolation before summing them to estimate the strength of a signal at a particular location. Advantages of the GTD are that it can consider diffraction and reflection and, further, that it can accommodate three-dimensional environments. In the shadow of a diffracting edge, the signal strength is inversely proportional to the diffracting angle. This means that, at grazing incidence, the GTD would predict an infinite signal strength (it is a 'singularity').The GTD produces predictions that are valid only away from the shadow boundary (the shadow-boundary region is also known as the 'transition region' between free space and deep in the shadow of the obstacle). Its advantage is that it produces very rapid predictions in complicated geometries. The uniform theory of diffraction (UTD) is a development of the GTD. It identifies a region around the shadow boundary where GTD predictions are inaccurate and uses a more accurate prediction, based on the Fresnel integral method, to compute the field strength in these regions. This can give the UTD the 'best of both worlds': accuracy in the transition regions and computational speed away from these regions. The UTD can also make accurate predictions when the diffracting edge is not evenly illuminated (such as in the case where one diffracting edge is in the shadow of another). It does this by introducing a secondary diffracting ray that is proportional to the gradient of the power density in the region of the diffracting edge. This produces a correction to account for the non-uniformity of the incident wave. The UTD is a well-established method of predicting the signal strength in complicated environments, such as those found inside buildings. Many software suppliers offer proprietary methods of predicting signal strengths that are based on the UTD.

3.10 Clearance requirements and Fresnel zones

Figure 3.2 shows that the diffraction loss equals zero when the Fresnel parameter equals about -0.8. Thus, if the link is planned so as to ensure that the Fresnel parameter associated with the most significant obstacle is no greater than -0.8, it is possible to calculate the losses on the link as if there were no obstacles. This is done in practice but microwave-link-planning engineers do not describe the clearance requirements in terms of the Fresnel parameter but, rather, in terms of a unit known as the 'Fresnel-zone radius'. The concept of the Fresnel zone relies on visualising an ellipsoid that surrounds the direct line joining the transmitter to the receiver on a microwave radio link. The ellipsoid is constructed so that the sum of the lengths of the straight lines joining the point on the ellipse to each end of the link is a constant distance greater than the length of the straight line between each end (the basic definition of an ellipse). The 'first Fresnel zone' is defined as the ellipsoid for which this additional distance exactly equals half a wavelength. In terms of the Fresnel parameter, if the apex of a knife-edge obstacle were located on the upper half of the ellipse, v would equal $\sqrt{2}$. The two-dimensional ellipse is shown in figure 3.20.

Referring to figure 3.20, the radius of the first Fresnel zone is given by

$$R = \sqrt{\frac{\lambda d_1 d_2}{d_1 + d_2}}, \tag{3.14}$$

where λ is the wavelength. All the parameters are in self-consistent units. If the radius is required in metres and the dimensions d_1 and d_2 are in kilometres and, instead of quoting wavelength, the frequency in MHz is used, then

$$R = 550\sqrt{\frac{d_1 d_2}{(d_1 + d_2)f}}. \tag{3.15}$$

The radius is a maximum at the mid-point of the path where $d_1 = d_2 = d/2$. Then

$$R_{\max} = 275\sqrt{\frac{d}{f}}. \tag{3.16}$$

We are interested in the clearance requirement. It is common to state this in terms of the fraction of the first Fresnel zone that must be free of obstructions. This is equated to the situation in which the Fresnel parameter is $v = -0.8$. A knife-edge obstacle with its apex at the lower edge of the first Fresnel zone would have a Fresnel parameter $v = -\sqrt{2}$. Now the value of the Fresnel parameter is proportional to the distance from the direct line. Since -0.8 is approximately 60% of $-\sqrt{2}$, if 60% of the first

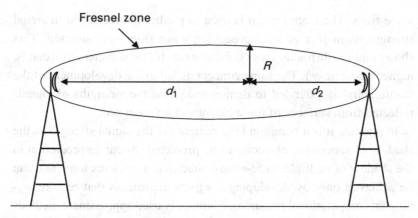

Figure 3.20 A two-dimensional view of the Fresnel ellipsoid.

Fresnel zone is free of obstructions diffraction loss can safely be ignored. This is the general rule applied by microwave-link planners.

As an example consider a microwave link operating at a frequency of 12 GHz over a distance of 25 km. The radius of the first Fresnel zone at the mid-point is given by

$$R = 275\sqrt{\frac{25}{12\,000}} = 12.6 \text{ metres.}$$

Thus clearance of at least 60% of 12.6 metres (7.6 metres) would be required at the mid-point. At a lower frequency, the radius of the Fresnel zone is larger. It is inversely proportional to the square root of the frequency. Therefore, at a frequency of 3 GHz, the Fresnel zone will be twice as big as at a frequency of 12 GHz. Thus, if the system described above operated at a frequency of 3 GHz, 15.2 metres of clearance would be required in order safely to ignore the effect of any obstacles.

3.11 Summary

Methods of predicting the strength of a diffracted signal in the shadow of obstacles have been discussed. The Cornu spiral has been introduced as a method of visualising the way in which the signal at a receiving point is the resultant of contributions from many wavelets on an approaching wave front. The Cornu spiral is used to predict the reduction in signal strength when part of an approaching wave front is obstructed. This shows that the diffraction loss is greater for shorter wavelengths (that is, higher frequencies). The same concept as led to the development of the Cornu spiral is extended to demonstrate how the strengths of signals reflected from surfaces of finite extent can be predicted.

In practice it is a common requirement for the signal strength in the shadow of successive obstacles to be predicted. Accurate prediction in the shadow of multiple knife-edge obstacles is a complex topic and can be achieved only by developing complex algorithms that require significant computational resources. Commonly used approximate methods of producing a prediction of the diffracted field in these circumstances have been explained and compared for a particular configuration. They all are seen to have their advantages and disadvantages.

From predictions of losses due to the presence of obstacles it is possible to develop clearance requirements such that, if no obstacles intrude within a given distance of the direct line between the transmitter and receiver, diffraction losses can be ignored. This involves the concept of the Fresnel zone, showing that the required clearance distance depends upon the distance from each end of a link together with the frequency of operation.

4   Reflection, scatter and penetration

When a radio wave can reach a receiver by more than one route, we say that the receiver is in a multipath environment. The way in which a standing wave pattern is established when the received signal is the combination of both a direct and a reflected signal is explained. The characteristics of the standing wave are shown to depend upon the nature of the reflection as determined by its reflection coefficient. Further examples of propagation paths involving reflection include propagation over a flat plane and propagation over water, the latter having the additional complication of tidal variation often causing the position of the reflecting surface to change. The more complex situation that arises when there are many different routes from transmitter to receiver is analysed. It is seen that the nature of the standing wave depends on whether one of the contributing paths is dominant (the 'Rician' environment) or whether the strengths of all of the signals on all paths are about equal (the 'Rayleigh' environment). It is further shown that the reflected signal depends on whether the reflecting surface is smooth or rough and the difference in the nature of the reflected signal is analysed. A further possible propagation mechanism is that of penetration of materials. The amount of penetration is seen to be dependent upon the electrical characteristics of the material and the frequency of the electromagnetic wave.

4.1 Introduction

In practical situations, radio waves will reflect off walls and off the ground. In mobile communications within cities, reflections off such surfaces often form the major propagation mechanism. When a radio wave reflects off a surface, the strength of the reflected wave is less than that of the incident wave. The ratio of the two is known as the 'reflection coefficient' of the surface. This depends upon the conductivity and permittivity of the material that forms the reflective surface, as well as its thickness and the frequency, angle of incidence and polarisation of the incident wave. Good conductors are near-perfect reflectors of electro-magnetic waves at frequencies used for radio communications. Dielectrics, such as glass, are less good reflectors, as are walls made of brick or con-crete. Another decision that has to be made when predicting the signal strength at a particular location is whether or not the surface can be regarded as smooth. When reflections are considered, the smoothness of the surface must be compared with the wavelength of the signal involved. Surfaces that appear, visibly (or to the touch), to be rough are often smooth to radio waves. Further, if the angle of incidence approaches 90 degrees, even rough surfaces can produce reflections as if they were smooth. This is because the roughness is a variation of surface height in a direction that is nearly at right angles to the direction of propagation. One phenomenon that occurs when you have a smooth reflecting surface is that of 'constructive' and 'destructive' interference. This is not interference in the commonly used sense but rather an interaction between the wave that travels to a point directly from a transmitter and a second wave from the same transmitter that undergoes a reflection on its way to the same point. The two waves will be coherent. That is, they will both look like perfect electromagnetic waves with exactly the same frequency. They will add on a phasor basis and will tend to add ('constructive' interference) and cancel out ('destructive' interference) at regular intervals in space, forming a standing wave pattern. The stronger the reflection the stronger the effect. The signal strength will vary very rapidly with distance and predicting the strength at a particular point is very difficult without very detailed information regarding the path geometry and the electrical characteristics of reflecting surfaces. Most signal-strength-prediction methods try to average out the peaks and troughs caused by the reflection to estimate a mean value.

When an electromagnetic wave is incident on a rough surface, the wave is not so much reflected as 'scattered'. The transition from coherent reflection to more random scattering is gradual as roughness increases. At large degrees of roughness (where the surface height varies by perhaps ten wavelengths or more) the actual strength of the scattered signal becomes much less than that of the incident signal even when the reflection coefficient of the material is very high. Thus, when the scattering surface is rough, the phenomenon by which the signal strength varies greatly over short distances is not as noticeable as it is when the surface is smooth.

The energy present in an incident radio wave that does not reflect from a surface must penetrate that surface. Thus, the reflection coefficient of the material affects the amount that penetrates into the material. Once inside the material, the wave will travel through the material. In most materials, the strength will decay as it travels. The electromagnetic wave heats up the material that it is propagating through: this is the principle by which microwave ovens function. The rate at which it decays depends upon the conductivity of the material: good insulators will cause little loss. Glass is an obvious example of a good insulator that allows electromagnetic waves (such as light) to propagate through it with little loss. An electromagnetic wave will not propagate through a good conductor at all before it is attenuated to negligible levels. A thin sheet of aluminium foil is sufficient effectively to 'stop' most electromagnetic waves (through a combination of reflection and attenuation). Even good insulators will exhibit some loss. This is due to the fact that, even though a current does not flow, the electric field causes electrical charges inside the molecules to vibrate, thus heating up the insulator (often referred to as a 'dielectric' in such cases). Coaxial cables become lossy at high radio frequencies because the dielectric spacing between the inner and outer conductors starts dissipating energy.

In planning practical radio systems, the penetration of typical building materials, such as brick and plaster board, by radio waves at various frequencies is of interest. Materials do not have to be continuous sheets to act as effective shields. A perforated sheet of metal will not allow any significant signal to pass through as long as the size of the holes in the metal is significantly smaller than a wavelength. This fact is exploited in the production of light-weight parabolic reflecting dishes for microwave frequencies.

4.2 Standing wave patterns

Just like light reflects from a mirror, so radio waves will reflect from a smooth surface. Since the wavelengths of radio waves are much longer than those of visible light, surfaces such as brick that appear rough and do not produce clear reflections at visible wavelengths will appear smooth to radio waves. When a surface is smooth you can predict the effect of the reflected wave by considering a virtual source on the other side of the reflecting surface. Radio waves tend to be narrow band (equivalent to a laser) and, when a direct and reflected wave interact, they form what is known as an 'interference pattern'. The word 'interference' can be confusing here. We are not talking about interference from an unwanted transmitter. Rather we are saying that the direct and reflected waves interfere with each other.

Imagine a receiver that receives a signal from a distant transmitter and a reflection of the signal from a smooth wall. The reflected signal has to travel further than the direct signal. This extra distance travelled means that the reflected signal will be out of phase with the direct signal by the time it reaches the receiver (this will be compounded by the fact that the wave will suffer a phase shift at the reflecting surface). At the receiver the direct and reflected waves will add together. To perform this addition we use a phasor diagram. This is very similar to a vector diagram that you use to add forces, except that each phasor represents a sine wave at a particular frequency with a particular phase shift. It is a fact that, if you add together two sine waves of exactly the same frequency with a phase offset, the result is another sine wave at that frequency with a phase different from that of either of the constituents. You can use a phasor diagram only if all the sine waves involved are of exactly the same frequency (they are then said to be 'coherent'). With reflected signals, the reflection will always be at exactly the same frequency as the direct signal (unless the reflection is from a moving object). Indeed, phasor diagrams are not limited to adding two sine waves together; as long as the signals are all coherent, you can use such a diagram to add together as many sine waves as you wish. For more on phasor addition, refer to the appendix.

We shall now return to our situation of 'direct signal plus single reflection' as illustrated in figure 4.1. Suppose that we locate our receiver at a spot where the direct and reflected signals add in phase. We should receive a strong signal. We shall now move the transmitter a fraction of a wavelength away from the reflecting surface towards the transmitter. The

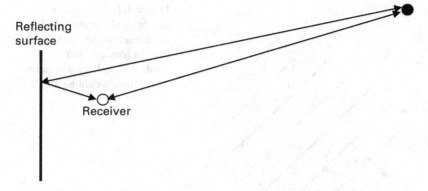

Figure 4.1 A plan view showing two possible routes for the signal to travel from transmitter to receiver: a direct route and via a single reflection.

direct signal will travel a shorter distance to the receiver and the reflected signal will travel a longer distance. The two signals will no longer be in phase and the resultant of combining the signals will not be as strong. A point will be reached where the two signals are in anti-phase. At this point the signal received will be a minimum. If the reflected and direct signals are of equal amplitude they will cancel each other out, resulting in zero signal being received. This can happen over relatively short distances. If the direct and reflected waves are travelling in exactly opposite directions at the receiver, a movement of only a quarter of a wavelength will change the situation from zero phase shift (a maximum) to a phase shift of 180 degrees (a minimum).

The direction of motion of the receiver is crucial. If the receiver moves along a line that bisects the angle between the direct and reflected waves then the phase difference will not alter and the signal level received will be constant. It is possible to draw a plan view with lines joining points of equal phase difference. The line joining points where the phase difference is zero will show where a maximum will occur and lines joining points where the phase difference is 180 degrees will show a minimum. Figure 4.2 shows a plan view of two coherent waves interacting. The solid black lines are the wave fronts representing the peaks of waves travelling in the direction indicated by the solid black arrow. The dashed lines represent the troughs of an electromagnetic wave travelling in the direction

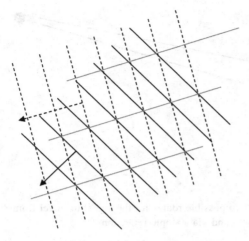

Figure 4.2 Wave fronts indicating the peaks (solid black line) of one wave and the troughs (dashed line) of another. Thin lines join points where there would be a null.

indicated by the dashed arrow. The faint lines join the positions where a trough of one wave will occur at the same location as a peak of the other. At these points there will be a null. As these waves travel, the points at which the null occurs stay in the same place and the interaction produces what is known as a 'standing wave' pattern.

In between the nulls there is a point where peaks from the two waves coincide. As the waves travel, troughs from the two waves will coincide at the same points as where peaks coincided a moment before. These points suffer a maximum disturbance. This clearly shows why we refer to an interference pattern being established between the direct and reflected waves. Figure 4.3 gives a three-dimensional view of the standing wave pattern. Rather than showing an instantaneous view of the electric field in a space, it shows the r.m.s. value of the electric field. The field would be varying at the frequency of the two coherent waves that produce the pattern.

One factor that affects the form of the interference pattern is the angle between the reflected and direct waves. We have used the situation where they are travelling in opposite directions. This produces the shortest distance between maxima and minima in the interference pattern. If the geometry of the situation is such as to produce small angles between the direct and incident waves, the distances will be much larger. The shortest distance between two successive maxima in an interference

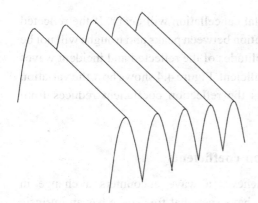

Figure 4.3 A three-dimensional view showing r.m.s. values of electric field strength within a standing wave pattern. Although the individual waves travel, the pattern remains stationary.

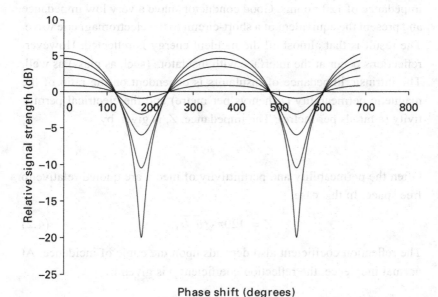

Figure 4.4 The variation in resultant signal strength for reflection coefficients of 0.9 (producing the sharpest troughs), 0.7, 0.5, 0.3 and 0.1 (producing the shallowest troughs).

pattern is $\lambda/(1 - \cos\theta)$, where θ is the angle between the directions of travel of the direct and reflected waves.

The standing wave patterns illustrated thus far are accurate for cases when the strength of the reflected wave is almost as strong as that of the incident wave. Then, near-total cancellation will occur. If the reflected signal is weaker then the variation between peaks and troughs will not be as great. The ratio of the amplitudes of the reflected and incident waves is known as the reflection coefficient. Figure 4.4 shows how the variation in signal strength reduces as the reflection coefficient reduces from 0.9 to 0.1.

4.3 Determining reflection coefficients

A reflection will occur whenever a wave encounters a change in impedance (a mismatch). We have seen that free space has an intrinsic impedance of 120π ohms. Good conductors have a very low impedance and present the equivalent of a short-circuit to the electromagnetic wave. The result is that almost all the incident energy is reflected. However, reflections occur at the interface with insulators (such as glass) as well. The intrinsic impedance of insulators is dependent on the ratio of the magnetic permeability (μ henrys per metre) and the electrical permittivity (ε farads per metre). The impedance, Z, is given by

$$Z = \sqrt{\mu/\varepsilon}. \tag{4.1}$$

Often the permeability and permittivity of media are quoted relative to free space. In this case,

$$Z = 120\pi\sqrt{\mu_r/\varepsilon_r}. \tag{4.2}$$

The reflection coefficient also depends upon the angle of incidence. At normal incidence, the reflection coefficient ρ is given by

$$\rho = \frac{Z - 120\pi}{Z + 120\pi}. \tag{4.3}$$

As an example, let us consider glass, for which $\mu_r = 1$ and $\varepsilon_r = 4$:

$$Z = 120\pi\sqrt{\mu_r/\varepsilon_r} = 120\pi/2 = 60\pi. \tag{4.4}$$

Therefore,

$$\rho = \frac{60\pi - 120\pi}{60\pi + 120\pi} = -\frac{1}{3}. \tag{4.5}$$

Thus the reflected wave will have an amplitude one third that of the incident wave. Thus a standing wave pattern similar to that shown in figure 4.4 for a reflection coefficient of 0.3 will be formed. The ratio of maximum to minimum signal strength is given by

$$\text{ratio} = 20 \log \left(\frac{1 + |\rho|}{1 - |\rho|} \right) \text{dB}. \tag{4.6}$$

Thus, if the reflection coefficient is 0.9, the ratio will be $20 \log(19) = 26$ dB and, for a reflection coefficient of 0.1, the ratio will be 1.7 dB. For the situation where the magnitude of the reflection coefficient is one third, the ratio of maximum to minimum signal strength will be about 6 dB.

The power density is proportional to the square of the amplitude and thus, if the reflection coefficient is one third, the power density of the reflected wave will be only one ninth that of the incident wave. A full treatment of this problem involves a detailed analysis of electromagnetic waves and boundary conditions that is beyond the scope of this book. For example, the reflection coefficient quoted above will be correct only when the incident wave is normally incident on the reflecting surface. Figure 4.5 shows the variation in reflection coefficient with angle of incidence for reflections over a horizontal plane with a relative permittivity equal to 4 both for vertical and for horizontal polarisation.

Notice that the reflection coefficients for the two polarisations are equal for an angle of incidence equal to zero. Further, notice that the magnitudes of the reflection coefficients for both polarisations are equal to unity for an angle of incidence equal to 90 degrees (grazing incidence). This illustrates the general rule that most surfaces have a very high reflectivity near to grazing incidence. Apart from the fact that the magnitudes of the reflection coefficients for the different polarisations are the same for angles of incidence of 0 degrees and 90 degrees, the two curves are very different. In particular, it can be seen that generally the reflection coefficient for vertical polarisation is less than that for horizontal polarisation and that there exists, for vertical polarisation, an angle of incidence for which the reflection coefficient equals zero. This angle is referred to as the Brewster angle.

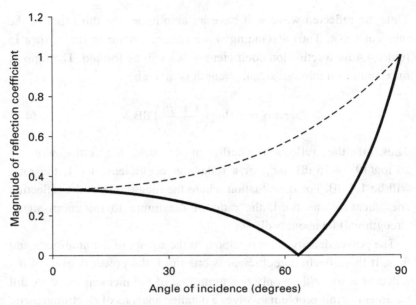

Figure 4.5 The variation of reflection coefficient with angle of incidence for vertical polarisation (solid line) and horizontal polarisation (dashed line), where the reflecting surface is a horizontal plane with a relative permittivity equal to four.

4.4 Propagation over a flat plane

Predicting the signal strength produced by a transmitter when it lies above a flat plane is a classical propagation problem. Even fairly rough ground of imperfect conductivity will reflect a very high proportion of the incident signal. Further, the wave will suffer a phase shift of 180 degrees on reflection (the reflection coefficient is almost equal to −1). Because the reflection coefficient is very high, there is a large variation between the peaks and troughs in the interference pattern produced. Because the angle between the direct signal and the reflected one decreases with distance from the transmitter, the distance between peaks and troughs also increases. Figure 4.6 shows the variation in signal with distance for assumed heights of both transmit antenna and receiving point of 40 metres at a frequency of 500 MHz.

Over short distances (up to about 10 km) the signal strength is seen to vary as points where alternately constructive and destructive

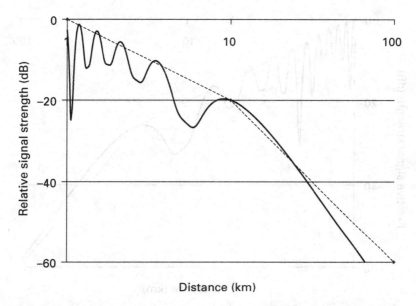

Figure 4.6 The variation of signal strength with distance over a smooth reflecting plane for transmitter and receiver heights of 40 m at a frequency of 500 MHz.

interference is experienced are passed through. The nulls can be very deep. At a particular distance (known as the 'break point') the oscillation in signal strength ceases. This is because the angle between the direct and reflected rays becomes so small that the first maximum is always above the height of the receiving antenna. As the distance increases further, the height of the first maximum continues to increase and the height of the antenna represents a smaller fraction of the distance from the ground to the maximum. Thus, as distance increases, two factors (free-space loss and the field strength at a fixed height representing a smaller fraction of the field strength at a maximum) combine to accelerate the increase of path loss with distance. The result is that the signal strength reduces faster than it does before the break point. At distances less than the break point the signal reduces with free-space loss but experiences both constructive and destructive interference so that the general trend of reduction in signal strength with distance follows a 20 log(distance) line. This trend is shown as a dashed line in figure 4.6 for distances less than 10 km. Beyond the break

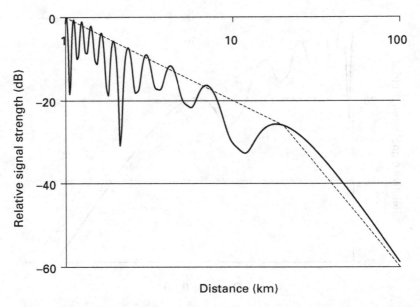

Figure 4.7 Signal strength against distance for a receiver height of 80 metres.

point at 10 km the signal strength will reduce not only as the free-space loss increases but also because the antenna is increasingly at a point of destructive interference. This causes the reduction in signal strength with distance beyond the break point to follow a trend closer to 40 log(distance). The dashed line in figure 4.6 shows indicates a 40 log(distance) slope at distances between 10 km and 100 km. If the height of the receiving antenna is increased to 80 metres, the distance to the break point also doubles, to 20 km, as indicated in figure 4.7.

If the frequency is reduced then the break point reduces proportionately. The distance to the break point can be estimated by using the equation

$$\text{break point} = \frac{4h_1h_2}{\lambda},\tag{4.7}$$

where h_1 and h_2 are the heights of the transmitting and receiving antennas, respectively, and λ is the wavelength. The equation is in 'self-consistent' units, meaning that, if the input parameters are all in metres, the result will also be in metres.

4.5 Propagation over water

Another example of the significance of reflections to radio-system plan-ners is that of planning a microwave link over a reflecting surface such as water. At the location of the receiver, reflections cause the signal to vary significantly with height. If the receiving antenna is located in a null, the communication link will fail. If the reflecting surface is water whose level varies with the level of the tide then the position of the null will also vary. This means that there is a good chance of any fixed receiving antenna being located in a null at some time in the tidal cycle. This is a real problem on microwave links. The only definite way of establishing the severity of the problem is by actually measuring the depths of the nulls. However, it is a usual requirement to design a microwave link without resorting to building a trial link. Sometimes, the electrical properties of the water are sampled and analysed and the reflection coefficient calculated, but a viable alternative is to plan as though the null will be very deep and build a receive system that will cope with this. A 'space-diversity' system is such a receive system. It is possible to build a receiver with two antennas such that you can almost guarantee that they will never both be located sim-ultaneously in a null. A crucial parameter in the design of such a system is the separation of the two antennas. A null is located where the direct and reflected waves are in anti-phase. The phase difference is dependent upon the relative path lengths and any phase shift at the reflecting surface. Figure 4.8 shows a 'path profile' for an over-sea radio link.

The reflected wave travels further than the direct wave, as shown in figure 4.9. The length of the reflected wave is taken as the length from the virtual reflection of the source. The separation between the real and virtual sources will be $2h_1$.

For small angles (note that the diagram exaggerates the height of the masts compared with the path length, making the angles between the direct and reflected paths appear much greater than they will be in practice), a null, or 'trough', will occur where

$$\frac{2\pi}{\lambda} 2h_1\theta + \phi = \left(n + \frac{1}{2}\right)2\pi, \tag{4.8}$$

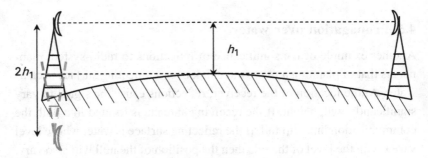

Figure 4.8 The path profile of a microwave link over water.

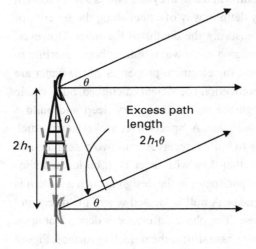

Figure 4.9 The excess path length as a function of angle.

where θ is in radians, n is an integer and ϕ is the phase shift at the reflection (also in radians). A peak will occur where

$$\frac{2\pi}{\lambda} 2h_1\theta + \phi = 2n\pi. \qquad (4.9)$$

The phase shift on reflection, ϕ, may be unknown. This means that it might not be possible to establish the exact height of a null or a peak (if the reflecting surface changes position, for example with the tide, this height will change anyway), but it is possible to estimate the separation between a peak and a null. If we construct a space-diversity system with antenna separation equal to the separation between a peak and a null, then the two antennas should never simultaneously be in a null.

Letting the angle θ at a peak be known as θ_{peak} and the angle at a null be known as θ_{null}, we get

$$\theta_{null} = \frac{(n + \frac{1}{2})2\pi - \phi}{4\pi h_1} \lambda,$$

$$\theta_{peak} = \frac{2n\pi - \phi}{4\pi h_1} \lambda,$$

$$\theta_{null} - \theta_{peak} = \frac{\lambda}{4h_1}. \tag{4.10}$$

At a distance d, the separation between a peak and a null will be approximately $\lambda d/(4h_1)$. We do not often think of the wavelength of a link but, rather, of the frequency. The relationship between wavelength in metres and frequency in MHz is given by

$$\lambda = \frac{300}{f} \tag{4.11}$$

and the separation, d_{sep}, between peak and null will be given by the approximation

$$d_{sep} \approx \frac{75d}{fh_1} \text{ metres}, \tag{4.12}$$

where f is in MHz and d and h_1 are in the same units (e.g. both in metres). It is more common to use kilometres as the unit for d. The same approximation is true for d in kilometres if f is quoted in GHz. Remember that h_1 is the height above the reflecting surface. This will vary with the tide height and, also, the position of the reflecting surface under consideration will change depending on the height of the receiving point. Nevertheless, this expression gives a good initial estimate of the separation between a peak and a null. For example, on a 7-GHz link of distance 14 km, using antennas of height 15 m (relative to the reflecting surface), a separation of 10 m would be expected. Note that you need not provide your antennas with this full amount of separation, the important issue is to ensure that both antennas are not simultaneously in a null (so a 10-m separation in this case could well prove unnecessarily expensive).

Even a few metres' separation would ensure that any simultaneous fade would be severely limited.

4.6 Rayleigh and Rician multipath environments

A terminal in a mobile radio network operates in environments where reflections are likely. It will move through an interference pattern whilst maintaining a connection with the base station. We say that the receiver must tolerate a 'multipath environment'. Many such environments will not be limited to a direct wave and a single reflection but may rather consist of many reflections. Nevertheless, the principle of all the waves adding to produce a single equivalent sine wave still holds. As the mobile moves in a particular direction, the phase from each of the signals will change by a different amount. This will affect the magnitude of the resultant wave. The interference patterns produced when there are many reflections will be much more complicated than for a single reflection. Nevertheless, at a fixed location (assuming that the reflecting surfaces and the transmitter do not move) the phase differences for all components will be constant. Thus the interference pattern itself does not move (it is a 'standing wave' pattern). However, unlike in the two-component case, it will not be possible to draw a straight line joining maxima or minima; rather the picture is now more like a mountain range with sudden peaks and dips.

When the receiver moves through such an environment we say that it experiences multipath fading. Initially, we shall consider a case where there are six components of similar amplitudes. As our starting point we shall assume the (very rare) situation in which all components are nearly in phase with each other. At this location the received signal will be a maximum (it is at the very peak of the mountain-range-like standing wave pattern). As we move in some random direction, the relative phases will change. The rate at which the phase changes for each of the components depends upon the angle between the direction of motion of the receiver and the direction of travel of the wave in question. If the receiver moves at right angles to the direction of the wave then the path length will not change at all. If the motion is along the line of the direction of the wave then the path length and hence the phase will change by a

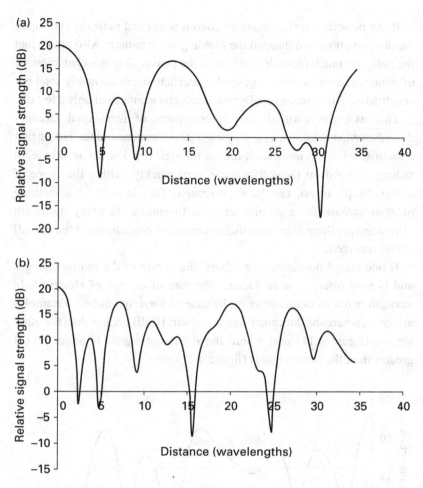

Figure 4.10 Variations in signal strength experienced by a receiver travelling in two different directions through an interference pattern produced by six nearly-equal components are shown in (a) and (b).

maximum amount of 2π radians for every wavelength moved. In general the phase of each component will change by $2\pi d \cos \theta/\lambda$ radians when the receiver moves by a distance d in a direction at angle θ to the direction of motion of the wave. Figure 4.10 shows possible variations in signal strength as the receiver travels in two random directions from a point near the maximum.

It can be seen that the distance between peaks and nulls depends upon the direction travelled through the standing wave pattern. Also notice that the nulls are much more dramatic than the peaks. This is a characteristic of radio wave interaction in general: cancellation can be nearly total but constructive interference usually produces enhancements of only a few dB.

This example with all the contributions of near-equal strength describes what is known as a 'Rayleigh' environment and the fading that a mobile will experience as it moves through it is known as 'Rayleigh fading'. Very deep fades that occur very quickly, within the space of a wavelength or so, can be experienced. This is seen as something of a worst case for a mobile terminal to endure. In many multipath environments there is an identifiable dominant contributor to the overall signal received.

If one signal dominates the others, the nature of the fading changes and is now referred to as Rician. The rate of change of electric field strength is not as dramatic as in the case of Rayleigh fading. Examples are given where the dominant signal is about 10 dB greater than the other signals (figure 4.11) and where the dominant signal is about 20 dB greater than the other signals (figure 4.12).

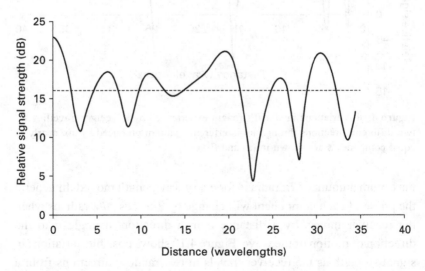

Figure 4.11 Multipath fading produced by six signals with one signal approximately 10 dB stronger than the other five.

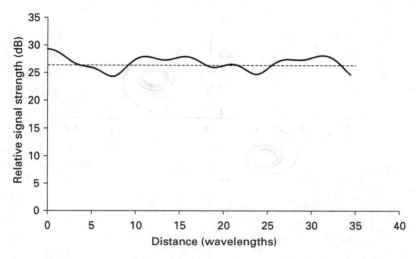

Figure 4.12 Multipath fading produced by six signals with one signal approximately 20 dB stronger than the other five.

It can be seen that the greater the dominance of the strongest signal, the less variation there is in the aggregate signal. Note that, in the case where there is 10 dB more power in the dominant signal, the amplitude of this signal is 3.2 times the average of the other five. Thus the other five signals could still cancel out the dominant signal, producing deep nulls. In the case where the dominance is 20 dB, the amplitude of the dominant signal is ten times that of the average of the other five and very deep nulls are an impossibility. In fact, the maximum variation between a peak and a null in this case would be less than 10 dB. There is no clear-cut change between the Rician and Rayleigh cases. The contributing signals will never be of exactly the same strength. Radio scientists refer to the 'degree' of a Rician environment or to the 'dominant–remainder' ratio. This ratio is the power of the strongest signal compared with the combined power of all the other signals. If the dominant signal is about 20 dB (a factor of 100) stronger than the other signals, but there are five other signals of about equal strength, then the dominant–remainder (or D–R) ratio will be about 20, or 13 dB.

4.7 Reflections from rough surfaces

Reflections from rougher surfaces cause different types of interference pattern. We shall use the Cornu spiral to help us once more. The Cornu

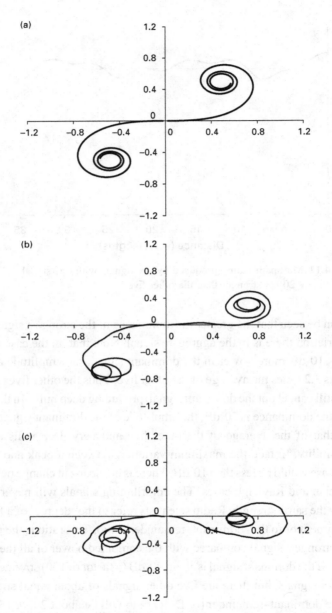

Figure 4.13 Distortion of the Cornu spiral for random phase variations of up to 0(a), 0.25(b), 0.5(c), 1.0(d), 3.0(e) and 10.0(f) radians between successive wavelets.

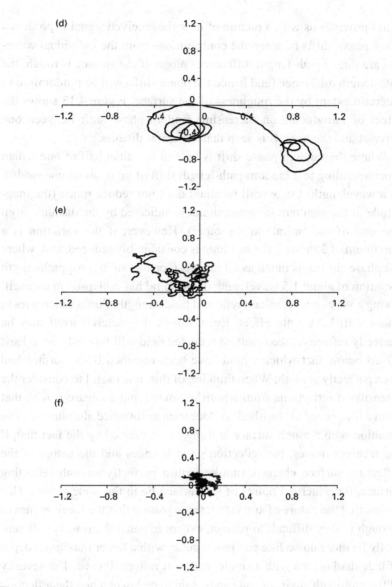

Figure 4.13 (*cont.*)

spiral provides us with a picture of how the received signal is produced when phase shifts between the contributions from the individual wavelets are due to path-length differences alone. If the surface is rough, the path-length difference (and hence the phase shift) will be randomised to a certain extent by the roughness of the surface. Figure 4.13 shows the effect of introducing an increasing random phase shift between one wavelet and the next. It is seen that the spiral distorts.

Where the random phase shift is kept to less than half of one radian (corresponding to a random path-length shift of up to about one twelfth of a wavelength) the overall resultant does not reduce much (the magnitude of the resultant reflected signal is indicated by the distance from one end of the 'spiral' to the other). However, if the variation is a maximum of 3 radians, the resultant is considerably reduced, and, where the phase shift is as much as 10 radians (corresponding to a path-length variation of about 1.5 wavelengths), the spiral has collapsed in on itself, giving a very small resultant reflected field strength. Further increases in phase shift have little effect. Even though the material itself may be perfectly reflective, the resultant reflected field will typically be at least 10 dB below that which would have been obtained if the surface had been perfectly smooth. When thinking of this, it is useful to consider the intensity of reflections from a polished metal and compare it with that from a roughened or 'brushed' surface such as 'brushed aluminium'. The situation with a rough surface is further complicated by the fact that, if the receiver moves, the reflection point changes and the nature of the reflecting surface changes (unlike with a perfectly smooth reflecting surface, for which all points of the surface lie in the same plane). This means that the nature of any interference pattern that the receiver moves through is very difficult to predict, except in general terms. It will generally be more noise-like and less regular, with a lower maximum depth of fade than occurs with a single, coherent reflected wave. The severity of the multipath environment can be said to be lower when the reflecting surface is rough. In reality, no surface is absolutely smooth or absolutely rough. In deciding whether the surface is rough or smooth for our purposes, we use what is known as the 'Rayleigh criterion'. This considers the geometry of the incident wave and the reflecting surface. The roughness of the surface is defined by its standard deviation, σ, relative to a perfect plane. A constant, C, is then derived:

$$C = \frac{4\pi\sigma \cos \Psi}{\lambda},$$ (4.13)

where Ψ is the angle of incidence.

Notice that the angle of incidence is a parameter. Even rough surfaces at near 'grazing' incidence appear to be quite smooth. If the surface is smooth you can expect peaks and troughs to form. If the surface is rough you do not get regular peaks and troughs but, rather, the interference pattern becomes more noise-like. There is, however, a transition region between the two, for which you will observe ever-weakening peaks and troughs as the roughness increases.

Generally, if $C < 0.1$ then the surface can be regarded as smooth, but if $C > 10$ then the surface is regarded as rough and the reflections will be diffuse. Between the values of 0.1 and 10 lies a transition region. In our experiments with the Cornu spiral a transition was certainly observed, but it is of interest to see how the value of C relates to values of variation of phase.

If an 'idealised' rough surface is considered to have two different levels, separated by a distance, d, as shown in figure 4.14, then one part of the wave will travel an extra distance. This extra distance equals $2d \cos \Psi$ and the phase difference, $\Delta\phi$, between these two parts is given by

$$\Delta\phi = \frac{4\pi d \cos \Psi}{\lambda}.$$ (4.14)

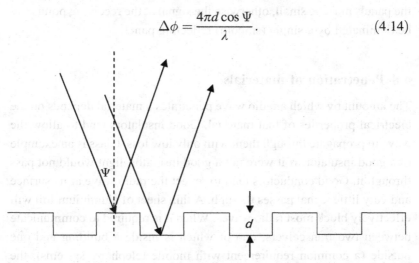

Figure 4.14 The path geometry for reflection from an idealised rough surface.

The similarity between this and the constant used for the Rayleigh criterion is clear. In the experiments with the Cornu spiral we assumed a random variation in phase between certain limits. The Rayleigh criterion assumes a normal distribution of surface heights with a standard deviation defined. However, our experiments with the Cornu spiral showed that if a random variation of ± 10 radians in phase existed then the resultant of the reflected signal collapsed to a negligible level compared with that of a reflection from a smooth surface. This tallies well with the standard deviation of 10 radians that is set as the threshold for regarding the surface as rough. Further, the Cornu spiral for a variation in phase limited to ± 0.1 radians still had the appearance of a good spiral and justifies the statement that the surface could be regarded as smooth. Whether it is better to consider random variations within a certain limit or a standard deviation of surface height is a matter of opinion. It should be borne in mind, however, that a normal distribution that is characterised by its standard distribution does stretch between $\pm \infty$.

The Rayleigh criterion necessarily involves a conversion from the roughness of a surface to the phase variation caused. Diagrams such as figure 4.14 that are used to support this argument often show the surface not to be rough as such but rather stepping between two levels and composed of smooth panels. For a surface to be rough, the size of each of the panels must be small, otherwise the signal at the receiving point may be dominated by a single (smooth) reflecting panel.

4.8 Penetration of materials

The amount by which a radio wave penetrates a material depends on the electrical properties of that material. Good insulators tend to allow the wave to propagate through them with only low loss. Glass is an example of a good insulator. If it were not a good insulator, light would not pass through it. Good conductors tend to reflect the radio wave at its surface and very little signal passes through. A thin sheet of aluminium foil will effectively block most radio waves. When it is required to communicate between two transceivers, one of which is inside a building and one outside (a common requirement with mobile telephony systems), the penetration loss of building materials becomes very important, as does the internal structure of the building. A building with an open-plan structure whose walls contain large glass windows will introduce little extra path loss (less than 5 dB), whereas a building with thick stone walls and small windows and an internal structure consisting of many solid walls can introduce extra path loss amounting to several tens of dB.

The process by which an electromagnetic wave travels from one side of a sheet of material to the other involves propagation (typically through a lossy medium) and multiple reflections.

The recommendation ITU-R P. 1238 publishes information that facilitates the calculation of typical values for the losses introduced by building materials. This information is published in the form of a material's 'complex permittivity'. The imaginary part of the complex permittivity can be converted into an effective resistivity that represents the lossy nature of the material. The loss depends upon the frequency: the higher the frequency the greater the loss. As an example, the permittivity of concrete is quoted as $7 - j0.85$ at a frequency of 1 GHz. This will allow us to estimate both the amount of energy in an incident wave that gets reflected at the surface and the amount that will be lost on passing through the concrete. A full treatment of this problem involves a detailed analysis of electromagnetic waves and boundary conditions that is beyond the scope of this book. However, it is possible to make deductions and approximations that provide a useful practical insight into the problem.

Firstly we introduce a little bit of arithmetic to make the problem simpler. We have stated that the reflection coefficient is given by

$$\rho = \frac{Z - 120\pi}{Z + 120\pi}. \tag{4.15}$$

Now if, as is usual, the material is non-ferrous ($\mu_r = 1$), it is possible to re-write the equation as

$$\rho = \frac{120\pi/\sqrt{\varepsilon} - 120\pi}{120\pi/\sqrt{\varepsilon} + 120\pi} = \frac{1 - \sqrt{\varepsilon}}{1 + \sqrt{\varepsilon}}. \tag{4.16}$$

Thus, for concrete, the reflection coefficient would be

$$\frac{1 - \sqrt{7 - j0.85}}{1 + \sqrt{7 - j0.85}} = -0.449 + j0.064 = 0.454\langle 172°. \qquad (4.17)$$

Thus the incident wave would be expected to undergo a near phase reversal on reflection and the amplitude of the reflected wave would be approximately 45% that of the incident wave. That means that the power density of the reflected wave would be only 20% that of the incident wave. By the law of conservation of energy the wave entering the concrete would have a power density 80% that of the incident wave. Once inside the concrete, the wave would propagate through but would dissipate energy as it travelled. The loss is determined by knowing the frequency and the imaginary part of the permittivity. The higher the frequency and the bigger the imaginary part of the permittivity, the greater the loss. For concrete with the electrical characteristics specified, the loss would be about 29 dB per metre at a frequency of 1 GHz. The loss is roughly proportional to frequency. So at 500 MHz the loss would be 14.5 dB per metre and at 2 GHz the loss would be 58 dB per metre.

4.9 Summary

We have seen that standing wave patterns form when two or more coherent waves interact. Where the two waves are in phase, a maximum disturbance will occur, whereas, where they are in anti-phase, a minimum or 'null' will occur. If the reflecting surfaces that cause the multipath situation do not move, the locations of the maxima and minima will not move, hence the name 'standing wave'. The depth of the null in a standing wave pattern is dependent upon the magnitude of the reflection coefficient of any reflecting surface. If the reflected wave is almost as strong as the direct wave then the difference between the signal strengths at maxima and minima will be great. The magnitude of the reflection coefficient can be determined if the electrical characteristics of the reflecting surface are known. Good, smooth, conductors will provide near-perfect reflection.

The effect of reflections is shown to be very noticeable when considering propagation over a flat plane. At near-grazing incidence the reflection coefficient, even of poor conductors, is very high and leads to strong peaks and nulls. Although the variation in signal strength with distance can be very high in this circumstance, the general trend is for the signal strength to reduce with distance. At a certain distance, the 'break point', the oscillation between peaks and nulls ceases and the signal strength reduces more rapidly with distance. The distance of the break point from the transmitter is seen to be dependent upon the heights of the transmitting and receiving antennas and the frequency of operation.

Microwave links propagating over water provide another example where the effect of reflections can be dramatic. Water forms a good smooth reflecting surface, particularly at near-grazing incidence. If the direct signal and that reflected from the water arrive at the receiving antenna in anti-phase, a null will occur and the link will fail. The best way of mitigating against this effect is to establish a second receiving antenna such that they will not both be in a null simultaneously. The most suitable separation distance between the two antennas has been derived and found to depend upon the transmitting antenna height, the path length and the frequency of operation.

Multipath environments have been characterised according to whether or not a single signal dominates. In a Rayleigh environment, all signals will be of about the same strength. This produces a very chaotic standing wave pattern with sharp nulls and peaks. If one of the con-tributing signals dominates then the environment is said to be Rician in nature. In this situation, the strength at any one point is dominated by the strongest signal, with the effect of the other signals resembling the effect of noise on a signal. The variation of signal strength with distance is not so great in a Rician environment as it is in a Rayleigh environment. The Cornu spiral has been used to demonstrate the effect of increasing the roughness of a surface. It is seen that the reflected signal strength, even from good conductors, is very small if the phase of the reflected signal varies randomly by 3 radians or more. The final section examines the way in which radio waves can penetrate obstacles. Good conductors will form a near-total block to radio waves, but electrical insulators can be lossy at high frequencies. Concrete was chosen as a typical example and the insertion loss was derived from its electrical characteristics.

5 Estimating the received signal strength in complex environments

A radio wave can often travel from a transmitter to a particular point by a number of routes: directly, by diffraction, by reflection, by penetration. At any point, the power received by a receiving antenna will be a combination of all these propagation mechanisms. Because the combined signal is a phasor sum of all the individual contributions an accurate prediction of the electric field strength is very difficult to obtain: it would need knowledge of the distance travelled for each propagation mechanism to within about a tenth of a wavelength plus details of the electrical properties of any materials involved in the paths involving reflection or penetration. Usually, all that is practical is to estimate the strength of the signal that would be achieved by each propagation path in isolation. The total received power is then estimated as the sum of these individual contributions. This gives an estimate of what is called the 'local-mean' level. That means that the actual power received would vary about this level by an amount that depends upon the relative strengths and directions of the individual contributions. If the angular separation of the individual contributions, when viewed from the receiver, is small then the signal will not vary very quickly with distance. Further, if one of the contributions is dominant and provides the majority of the signal power on its own then the variation will not be as great as if all the different propagation paths contributed nearly equal amounts of power.

5.1 Aggregating multiple contributions

In most practical situations, a mobile hand-held receiver will find itself in a multipath environment. One possible method of producing an estimate of the resultant signal at a point in a multipath environment is to add the powers of each component. Figure 5.1 shows the effect of this for a Rician

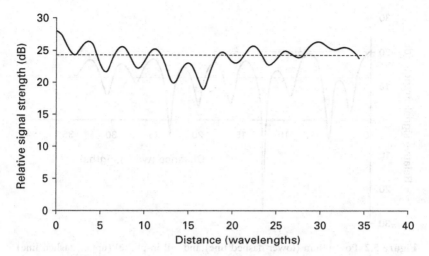

Figure 5.1 A power-sum estimate (dashed line) of signal strength in a Rician environment.

environment with a dominant–remainder ratio of 13 dB. The dashed line shows the sum of the powers of the individual elements. Notice that the signal strength at a point could be above or below this level. However, for the Rician environment with a single dominant signal, the power sum provides a reasonable estimate. The power sum is a single figure since the individual contributions are constant; the resultant field strength changes because the relative phases of these contributions change.

If we examine the prediction for the Rayleigh environment (see figure 5.2), the dashed line representing the power sum is still probably as good an estimate as any, but the large variation of the signal strength with position means that the estimate is less reliable. If we wish to predict the largest possible value of the signal, we can make a prediction assuming that all the elements add in phase with each other. This produces the higher dashed line on the diagram. This method is used occasionally if a conservative estimate of the level of an interfering signal is required. The fact that the actual signal at a point may be several dB above the power sum would be a cause of concern in that situation.

As an example, let us consider the situation depicted in figure 5.3. The receiver has no direct line of sight to the transmitter. The received signal

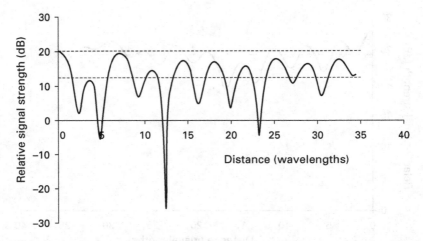

Figure 5.2 Power-sum (lower dashed line) and 'all-in-phase' (upper dashed line) estimates of signal strength in a Rayleigh environment.

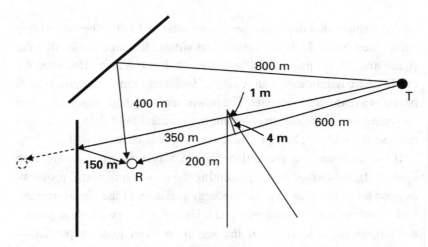

Figure 5.3 A path geometry showing multiple propagation paths from transmitter (T) to receiver (R).

is deemed to be the sum of three major contributors: a signal diffracted around the corner of a building; a signal reflected off a building some distance away; and a signal that has undergone both diffraction and reflection. The distances of each path are marked on the diagram. The frequency is deemed to be 500 MHz and the transmitter and receiver both have omni-directional radiation patterns. The transmitting antenna emits an EIRP of 20 watts (43 dBm). The magnitude of the reflection coefficient at both surfaces is estimated to equal 0.3 (the electric field strength reflection coefficient).

We shall consider first the diffraction path. The line-of-sight distance is 800 metres. The free-space receive level would be $43 - 32.4 - 20 \log (0.8) - 20 \log(500) = -41.4$ dBm.

Diffraction loss can be estimated by calculating the value of the Fresnel parameter v from

$$v = h\sqrt{\frac{2}{\lambda}\left(\frac{1}{d_1} + \frac{1}{d_2}\right)}. \tag{5.1}$$

The wavelength $\lambda = 0.6$ metres, $d_1 = 600$ metres and $d_2 = 200$ metres. The amount by which the line of sight is obstructed, h, equals 4 metres. Thus

$$\begin{aligned}
v &= h\sqrt{\frac{2}{\lambda}\left(\frac{1}{d_1} + \frac{1}{d_2}\right)} \\
&= 4\sqrt{\frac{2}{0.6}\left(\frac{1}{600} + \frac{1}{200}\right)} \\
&= 0.596.
\end{aligned} \tag{5.2}$$

The loss that this would cause, assuming knife-edge diffraction, is given by

$$\begin{aligned}
\text{loss} &= 6.9 + 20 \log\left(\sqrt{(v - 0.1)^2 + 1} + v - 0.1\right) \\
&= 6.9 + 20 \log\left(\sqrt{(0.596 - 0.1)^2 + 1} + 0.596 - 0.1\right) \\
&= 11 \text{ dB}.
\end{aligned} \tag{5.3}$$

Thus, the received signal strength, if this were the only propagation path, is $-41.4 - 11 = -52.4$ dBm.

We now turn to the reflected path. The total path length is 1200 metres. The free-space received signal strength would be $43 - 32.4 - 20\log(1.2) - 20\log(500) = -45.0$ dBm. The reflection coefficient is 0.3. The reflection loss is $-20\log(0.3) = 10.5$ dB, thus the contribution from this path is $-45.0 - 10.5 = -55.5$ dBm.

The third path can be analysed as though the route of the signal were from the transmitter to the virtual image of the receiver. The direct distance is 1100 metres, giving a free-space receive level of -44.2 dBm. The reflection loss is 10.5 dB, making the power of this reflected signal in the absence of any obstacle -54.7 dBm. Now to calculate the diffraction loss. The value of d_1 is 600 metres as before, but the reflection has led to an increase in the value of d_2 to 500 metres. The amount by which the path is obstructed is now only 1 metre, giving a Fresnel parameter of

$$v = 1\sqrt{\frac{2}{0.6}\left(\frac{1}{600} + \frac{1}{500}\right)}$$

$$= 0.111. \qquad (5.4)$$

This leads to a predicted diffraction loss of 7.0 dB, leading to an estimated receive power for this path on its own of $(-54.7 - 7.0) = -61.7$ dBm.

Thus the three most significant components are

- a diffracted signal producing -52.4 dBm of received power;
- a reflected signal producing -55.5 dBm of received power; and
- a signal undergoing reflection and diffraction, producing -61.7 dBm.

In order to compute the sum of these three contributions, it is necessary to convert them each into milliwatts:

Power in dBm	Power in milliwatts
-52.4	5.75×10^{-6}
-55.5	2.82×10^{-6}
-61.7	0.68×10^{-6}
Sum of powers	9.25×10^{-6}
Convert to dBm	-50.3 dBm

Thus an estimate of the received power would be -50.3 dBm, which is about 2 dB above the strength of the single strongest contributor.

Remember that the three signals will be coherent and will form an interference pattern with peaks and nulls. To gain an idea of the level of variation, we need to look at the strengths of the electric fields that the different contributing powers represent. We know that P_r (dBm) $=$ E (dBμV/m) $- 20 \log f$ (MHz) $- 77.2$. Thus E (dBμV/m) $= P_r$ (dBm) $+ 20 \log f$ (MHz) $+ 77.2$.

Since the frequency is 500 MHz, $20 \log f = 54.0$. This allows us to write the equation as E (dBμV/m) $= P_r$ (dBm) $+ 131.2$. For the signal powers received we can construct the following table:

Power (dBm)	Field strength (dBμV/m)	Field strength (μV/m)
-52.4	78.8	8710
-55.5	75.7	6100
-61.7	69.5	2990
	Sum of fields	17800
	Convert to dBμV/m	85.0
	Convert to power, dBm	-46.2 dBm

Thus we can see that the peak value of the signal strength, if all the electric fields happened to add in phase, would be -46.2 dBm (some 4 dB higher than the power sum). Since no signal dominates to the extent that its electric field is greater than the other two combined, it is possible for the phase relationship to occur where near cancellation is achieved and we can therefore expect some deep nulls to be formed, similar to those shown in the diagram for Rayleigh fading (figure 5.2).

6 Atmospheric effects

Atmospheric effects add significant complicating factors to the job of the radio-system designer. Initially, the way in which multipath propagation can be established due to the atmospheric structure is explained together with the effect this can have on the received signal. Another phenomenon, ducting, whose existence depends upon the structure of the atmosphere, is then discussed. It is seen that ducting can lead to levels of long-distance interference rising. The way in which diversity techniques can reduce the effect of multipath fading is explained. Another form of fading ('diffraction fading') is then described. Diffraction fading occurs when the atmosphere causes the path of the radio wave to bend upwards as it travels, leading to no line of sight existing between points for which a clear line of sight would be expected. Next, it is shown that, in severe cases of multipath propagation, the delay between two paths can be significant. This leads to the received spectrum in large-bandwidth links becoming distorted. The amount of fading caused by rain is explained as another factor that must be considered when designing a microwave radio link. Further, the fact that, even if no fading occurs, the ever-present molecules present in the atmosphere will cause attenuation is described, together with an indication of the frequency dependence of this attenuation. Finally, the way in which atmospheric losses affect the noise performance of radio systems, particularly Earth–space systems, is analysed.

6.1 Multipath propagation in the atmosphere

Microwave links using parabolic dish antennas mounted on tall towers usually rely on having a line of sight from the transmitter to the receiver. This means that it should be possible to see (on a clear day with the aid of binoculars) one antenna from the other. This fact may lead to an expectation that the received signal should remain at a steady level, particularly when the air is clear. This is not the case. At least, it is not the case for a large enough percentage of the time to make radio communication as reliable as cable. The Earth's atmosphere is not 'free space'. That, in itself, is not a problem. What causes a problem is the fact that the Earth's atmosphere is not homogeneous. Variations in pressure, humidity and temperature all lead to variations in refractive index. The atmosphere tends to form layers such that it is possible to identify a vertical refractive-index profile. It sometimes happens that more than one significant route is established from the transmitter to the receiver. The resultant power will be the phasor sum of all contributions. It is possible for the contributions to add in anti-phase and nearly cancel each other out. This situation represents a deep 'fade'. If the link was not designed with sufficient fade margin, then the received signal will drop below the level necessary to establish a satisfactory bit error ratio and an 'outage' is said to occur. A link will typically be designed so that such outages are limited to less than 0.01% of the time (or to no more than about 50 minutes per year). The likelihood of a fade occurring increases dramatically with link length (it is approximately directly proportional to the cube of the link length). It is also dependent on frequency, height above sea level and whether the path is horizontal or sloping (sloping, or 'slant', paths suffer less multipath fading). Further, certain parts of the world suffer more fading than others, with hot areas tending to fare worse than cooler areas. This is accounted for by determining a climatic factor for the location of the link. On a long link of a few tens of kilometres, it is common for fade margins of more than 30 dB to be needed in order to maintain satisfactory performance.

6.2 Ducting

A duct is something that will confine whatever is travelling along it into a narrow 'pipe'. The atmosphere can assume a structure that will produce a similar effect on radio waves. When a radio wave enters a duct it can travel with low loss over great distances. This is an issue when predicting interference levels at a great distance. An 'elevated duct' can occur when the refractive-index gradient reverses and starts increasing with height over a small interval. This will lead to a plane of high refractive index being sandwiched between regions of lower refractive index. The atmosphere will then act in the manner of a giant optical fibre, trapping the radio wave within the layer of high refractive index. A wave trapped in a duct can travel beyond the radio horizon with very little loss, producing signal levels within a few dB of the free-space level.

Another type of duct that does not require a reversal of refractive-index gradient occurs when this gradient is very steep. When this occurs, the radio wave can be bent towards the surface of the Earth, from which it will reflect, propagate upwards, become bent towards the Earth's surface again and so on. The loss with distance when such a duct exists is dependent upon the electrical properties of the surface, together with its roughness (the rougher the lossier because the reflection is not specular).

When a radio wave travels over the sea, it may be affected by an evaporation duct. The fast-changing humidity levels immediately above the surface of the sea cause the refractive index to reduce rapidly with height. Because of the prevalence of these ducts over water, and the fact that water provides a good, flat reflecting surface, trans-horizon interference is much more common on radio paths that travel over water.

When an interfering signal is trapped in a duct, it can have an effect hundreds of kilometres away. The interfering signal is said to be elevated. The statistics of elevation of interfering signals can be estimated using ITU-R P. 452. This recommendation considers all significant methods of radio wave propagation, not only ducting. However, short-duration increases in signal strength at large distances are usually caused by ducting.

6.3 The *k*-factor and the standard atmosphere

The Earth has an atmosphere. This means that terrestrial radio waves do not propagate in free space. Further, and this is what causes the problems for radio-planning engineers, the structure of the atmosphere is not uniform. Even on the least interesting days there is a general decrease in pressure and temperature with height.

Additionally, the Earth is not flat. It is very nearly a sphere. This means that a line joining a transmitter to a receiver will vary in height

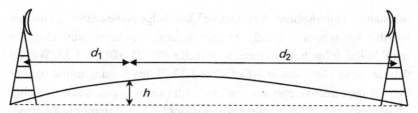

Figure 6.1 A path profile showing Earth bulge.

above sea level even if the transmitter and receiver are of equal height. Figure 6.1 illustrates this. We talk of the degree of 'earth bulge'.

This can be calculated and, at a given point, d_1 km from one end of a link and d_2 km from the other end, the height, h, of the Earth bulge is given in metres. It is at a maximum in the middle of the path and the maximum value of Earth bulge at the centre of a path of length d kilometres is approximately $0.02d^2$ metres. Thus, for a link of length 7 km, the earth bulge is only about 1 metre at the mid-point, but for a link of length 50 km the earth bulge will be 50 metres, which could be very significant. Suppose that we had a 30-km link with antennas of height 20 metres at each end. The Earth bulge in the centre of the link will be about 14 m and therefore the line joining the transmitter to the receiver will be only about 6 m above the ground at this point. The major implication is that the radio wave will pass through the atmosphere at varying heights above the ground and thus can be expected to encounter variations in temperature, pressure and humidity. These three parameters of the atmosphere affect the speed at which a radio wave will travel (we say that the 'refractive index' of the atmosphere changes). The effect is very small in terms of percentage but can have a significant affect over the length of the path. As the radio wave travels through atmosphere of varying refractive index it will tend to bend, rather than travel in a straight line. If the atmospheric parameters of pressure, temperature and humidity adopt a 'standard' distribution with height, then this actually helps the radio planner. The bending of the direction of travel will be very slight but will be in the same sense as the curvature of the Earth. Thus a radio wave will tend to follow the Earth's surface. The overall result is that the radio horizon is further than the visible horizon. The effect is the same as if the Earth bulge were reduced.

In a standard atmosphere, the effective Earth bulge at the centre of a path of length d km is approximately $0.015d^2$ metres: only three quarters of the 'real' bulge. It has been reduced by a factor of $0.02/0.015 = 1.33$. We say that the atmosphere has a 'k-factor' of 1.33. When making initial predictions of the effective curvature of the Earth (for example, when assessing the necessary heights for antennas), a factor of 1.33 is commonly assumed.

6.4 Anomalous propagation and multipath fading

For 90% or so of the time, the atmosphere is fairly well-behaved and the received signal is of a nearly constant strength. However, in designing point-to-point links in particular, unavailabilities of significantly less than 10% (typically 0.01%) are sought. For time percentages this low, the atmosphere can, and does, cause unusual things to happen. Of particular importance is the refractive-index gradient. If the refractive index steadily decreases with height (as it does in the case of the standard atmosphere) then the radio wave will tend to follow a curved path. If the gradient becomes very steep, the path of the radio wave will curve so much that it will be bent back down to Earth. It will reflect off the surface of the Earth and the process is repeated. In this way, the wave can propagate over long distances. We say that the atmosphere has formed a 'duct' and that what is occurring is 'ducting'. It can lead to problems with interference at large distances. If the refractive-index gradient varies with height, it can lead to more than one path being established between the transmitter and receiver. It is likely for the amplitudes of the signal from each path to be very nearly equal. If these signals arrive in anti-phase, cancellation can result (this is similar to the case in which reflection off the surface of water was discussed). This is referred to as a multipath fade and is a subject of much study. It is the most significant cause of outages on microwave links operating at frequencies below about 10 GHz (above about 10 GHz, fading due to rain is usually the major cause).

Being able to predict the probability of a multipath fade for a particular link is a necessary part of being able to design a microwave radio link. The major reference source for this is ITU-R P. 530. Recommendation P. 530 offers different equations depending on whether a 'detailed link design' or a 'quick planning application' is required. We shall illustrate the principle using the formulas for a quick planning application. The equation recommended for use in predicting the probability, p_w, of a fade exceeding depth A dB for small time percentages is

$$p_w = Kd^3(1 + |\varepsilon_p|)^{-1.2} \times 10^{-0.033f - 0.001h_L - A/10}, \qquad (6.1)$$

where d is the path length in km, f is the frequency in GHz, ε_p is the slope of the path in milliradians and h_L is the height of the lower of the two terminals above sea level. Examining the equation allows generalisations to be made regarding the dependence of fading on particular path parameters.

There is a large dependence on distance. The probability of a fade of a particular depth increases with the cube of distance. Thus, as the distance is doubled, the probability of a particular fade depth increases by a factor of eight. Or, alternatively, the fade for a given probability increases by 9 dB. So, doubling the distance will increase the free-space loss by 6 dB, and increase the probability of fading by 9 dB, thus increasing the overall link-budget loss by 15 dB.

There is a slight dependence on frequency. Increasing the frequency by 1 GHz will decrease the probability of a fade by a factor of 1.08.

There is a fairly strong dependence on the height of the path above sea level. There is simply less atmosphere at higher altitudes and therefore the effect of atmospheric fading is smaller. This is indicated in the equation by the height of the lower of the two terminals. For every 1000-metre increase in altitude the required fade margin reduces by 10 dB.

There is a dependence on whether the path is sloping or horizontal. A sloping path will be less susceptible to the effects caused by the horizontal layering of the atmosphere. On a 10-km path, if there is a 100-metre difference between the altitude of the two ends, 12 dB less fade margin will be required compared with the situation in which the two terminals are the same height above sea level.

Finally, notice the dependence of the probability on the actual fade depth, A dB. For an increase in fade depth of 10 dB the probability reduces by a factor of 10. Thus, if an availability of 99.99% is required, an extra 10 dB fade margin will be needed compared with an availability of 99.9%.

Some of these parameters (such as path length and frequency) will immediately be known. However, the probability of fading does change with location. This is accommodated for in the equation by the 'radio climatic factor', K. To estimate the value of K, a database or map giving information regarding the normal refractive-index gradient is needed. Again, the ITU can provide information on this (recommendation ITU-R P.453 contains a map, and data in tabular form can be obtained from the ITU).

The lower the value of the climatic factor the better insofar as multipath fading is concerned. Following the formula given in ITU-R P.530 for quick planning applications yields values that vary from about 1.2×10^{-4} in areas such as the southern oceans to as much as 14×10^{-4} around the Mediterranean sea. A typical value over much of northern Europe is 2.4×10^{-4}. This means that a microwave link in the Mediterranean would need about 8 dB more fade margin than an otherwise identical link in northern Europe for the same performance.

Once all the parameter values have been identified, it is possible to predict the necessary 'fade margin' that must be designed in to accommodate a fade with a particular probability. The process is as follows: ascertain the required availability and convert to an acceptable probability of a fade-induced outage; determine the depth of fade that corresponds to this probability; incorporate this fade margin within the power budget for the link.

Example: A 7-GHz microwave link carrying an 8-Mbit/s digital radio channel is to be established over a Mediterranean path. The minimum required signal level is -84 dBm. The path length is 35 km. Both terminals are at sea level but the antennas are at the top of 50-metre masts that have been installed at each end. Determine the required fade margin for an availability of 99.99% and, further, suggest appropriate transmit powers and antenna sizes assuming miscellaneous feeder and connector losses of 8 dB.

Step 1: use all available information to produce an equation linking the probability of fade to the depth of the fade in as straightforward a manner as possible. Inserting known values for parameters into the formula

$$p_w = Kd^3 (1 + |\varepsilon_p|)^{-1.2} \times 10^{-0.033f - 0.001h_L - A/10} \tag{6.2}$$

gives

$$p_w = 14 \times 10^{-4} \times 35^3 \times 10^{-0.033(7) - 0.001(50) - A/10}, \tag{6.3}$$

$$p_w = 31 \times 10^{-A/10}, \tag{6.4}$$

$$A = -10 \log\!\left(\frac{p_w}{31}\right). \tag{6.5}$$

Since the required availability is 99.99%, $p_w = 0.01$,

$$A = -10 \log\!\left(\frac{0.01}{31}\right) = 35\,\text{dB}. \tag{6.6}$$

Since the minimum required signal level is -84 dBm, the received signal level under non-fading conditions must be -49 dBm. At a frequency of 7 GHz, the free-space path loss over a distance of 35 km would be

$$92.4 + 20 \log(7) + 20 \log(35) = 140\,\text{dB}. \tag{6.7}$$

Adding the 8 dB feeder and coupler losses to this gives total system losses of 148 dB. Thus

$$P_t + G_t + G_r - 148 = -49\,\text{dBm}, \tag{6.8}$$

where P_t is the transmit power in dBm and G_t and G_r are the gains for the transmit and receive antennas, respectively, in dBi. The designer must decide on how to achieve the necessary receive power as a combination of transmit power and antenna gains. A typical antenna of diameter 1.2 metres has a gain, at 7 GHz, of

$$18 + 20 \log(1.2) + 20 \log(7) = 36\,\text{dBi}. \tag{6.9}$$

If such antennas were to be deployed, a transmit power of $+27$ dBm (0.5 watts) would provide the required signal level.

Further example: repeat the design problem assuming that the location has shifted to a mountainous area of northern Europe. Here the value of the climatic factor, K, is reduced to 2.4×10^{-4}. We shall assume that the lower antenna is at an altitude of 500 metres and the higher antenna is at an altitude of 600 metres. The link length remains 35 km and the frequency to be used is again 7 GHz. Miscellaneous losses are again assumed to total 8 dB. Since the difference in altitude between the two ends of the link is 100 metres, the value for ε is $100/35 = 2.9$. The probability of fading is given by

$$p_w = 2.4 \times 10^{-4} \times 35^3 \times (1 + 2.9)^{-1.2} \times 10^{-0.033(7) - 0.001(500) - A/10},$$

(6.10)

$$p_w = 0.37 \times 10^{-A/10}.$$

(6.11)

Thus

$$A = -10 \log\left(\frac{p_w}{0.37}\right),$$

(6.12)

$$A = -10 \log\left(\frac{0.01}{0.37}\right),$$

(6.13)

$$A = 15 \, \text{dB}.$$

(6.14)

The change of location has reduced the necessary fade margin from 35 dB to 15 dB. Thus the antenna gains and the transmit power can be reduced. If the antenna diameter is reduced from 1.2 metres to 0.6 metres, their gain will be reduced by 6 dB, thus decreasing the system gain by 12 dB. If the transmit power were then reduced by 8 dB to 19 dBm, the required performance would again be established.

It should be noted that the equations quoted from ITU-R P. 530 predict the unavailability for what is known as the 'worst month'. The concept of the worst month is well established as a suitably long time period over which very poor performance should not be tolerated. Sometimes requirements are quoted for the average worst-month distribution and sometimes for the average annual distribution. The ITU recommendation P. 530 gives guidance on converting from one distribution to the other.

6.5 Diversity techniques

The examples given above did not result in any unachievable requirements for the system. The antenna sizes and transmit powers stated are readily available. If the path length is increased to 60 kilometres and the capacity of the system to 140 MBit/s, the challenge becomes more severe. Let us consider the possibility of a requirement to establish a 140-Mbit/s, 60-km link in northern Europe between antennas 150 metres above sea level. Since the link is part of a network that consists of many such links, the required availability for any one link is 99.998%. Further, because the masts at each end are so high, losses in the feeder system now add to 22 dB.

Again using the equation from ITU-R P. 530:

$$p_w = 2.4 \times 10^{-4} \times 60^3 \times 10^{-0.033(7)-0.001(150)-A/10}, \qquad (6.15)$$

$$p_w = 21.4 \times 10^{-A/10}. \qquad (6.16)$$

Thus

$$A = -10 \log\left(\frac{p_w}{21.4}\right). \qquad (6.17)$$

For a required availability of 99.998%, $P_w = 0.002$. Therefore, the required fade margin is

$$A = -10 \log\left(\frac{0.002}{21.4}\right) = 40 \, \text{dB}. \qquad (6.18)$$

The free-space loss over a distance of 60 km at a frequency of 7 GHz equals 145 dB. The minimum required signal level for a capacity of 140 Mbit/s equals approximately $-154 + 10 \log(140 \times 10^6) = -72$ dBm. Therefore, the system design must deliver a signal level under non-fading conditions of -32 dBm. We can now determine the sum of the transmit powers and antenna gains:

$$P_t + G_t + G_r - 145 - 22 = -32, \qquad (6.19)$$

$$P_t + G_t + G_r = 135. \qquad (6.20)$$

Thus, if the transmit power of 27 dBm is used, antenna gains of 54 dBi at both transmit and receive ends will be required. At 7 GHz, the required antenna size is given by

$$\text{gain} = 18 + 20\log D + 20\log f, \tag{6.21}$$

$$\text{gain} = 18 + 20\log(Df), \tag{6.22}$$

$$D = \frac{10^{(\text{gain}-18)/20}}{f} = \frac{10^{(54-18)/20}}{7} = 9.0 \text{ metres}. \tag{6.23}$$

Deploying a 9-metre antenna (that would have a beamwidth of about one third of a degree) at the top of a mast more than 100 metres in height is not a realistic option. Increasing the transmit power and reducing the antenna size may be an option, but, even if a 10-watt transmitter were deployed, 4.5-metre antennas would still be required. A more cost-effective option would be to deploy a second receive antenna that would allow for a lower fade margin whilst maintaining the required availability.

6.5.1 Space diversity

In order for a deep multipath fade to occur, two near-equal signals need to add in anti-phase (they can occur with more than two signals, but we will stick to the simplest case). This can be thought of as an unlucky occurrence where the receiving antenna is in exactly the 'wrong' place. One method of reducing the likelihood of multipath fading is by using two receive antennas and using a switch to select the better signal. If these are physically separated then the probability of a deep fade occurring simultaneously at both of these antennas is significantly reduced. This is known as 'space diversity'. Recommendation ITU-R P. 530 gives information on how to predict the 'improvement factor' by which the predicted fade probability is reduced. Suppose that, in the previous problem, the largest antennas that could sensibly be deployed would be 3.6 metres in diameter and the maximum power 1 watt (3 dB more than the current 27 dBm). The gain follows a $20\log D$ form and therefore the gain of each antenna would be $20\log(9/3.6) = 8$ dB lower than needed (total system gains 16 dB too low). Thus the fade margin, instead of being the required 40 dB, would be only 24 dB. The extra transmit power would increase the fade margin to 27 dB. We now need to turn to ITU-R P. 530 for guidance on a necessary antenna spacing. This recommendation introduces a new term p_0, where

$$p_0 = Kd^3(1 + |\varepsilon_p|)^{-1.2} \times 10^{-0.033f - 0.001h_L}. \tag{6.24}$$

This term has actually come into previous calculations and, for the parameters of the link being studied here,

$$p_0 = 21.4. \tag{6.25}$$

Then, ITU-R P. 530 states that the improvement factor, I, for the link under study can be obtained from

$$I = \left[1 - \exp\left(-0.04 \times S^{0.87}f^{-0.12}d^{0.48}p_0^{-1.04}\right)\right] 10^{A/10}, \tag{6.26}$$

where S is the antenna spacing in metres and A is the fade margin without diversity (27 dB in this case). We need an improvement factor to compensate for the difference between the necessary fade margin of 40 dB and the actual fade margin of 27 dB. This difference of 13 dB transforms into a ratio of 20. Thus we need an improvement factor of 20. Figure 6.2 shows a graph relating the improvement factor to the antenna separation for the parameters of this link under investigation.

It can be seen from the graph that an antenna separation of approximately 6 metres will provide the necessary improvement factor. Summarising, increasing the power by 3 dB and using two receiving antennas of diameter 3.6 metres separated by 6 metres will give the same availability as the original design with antennas of diameter 9 metres.

6.5.2 Frequency diversity

Another method of diversity uses two separate carrier frequencies ('frequency diversity'). The fade is caused by the path-length difference between the two paths being $(n + \frac{1}{2})\lambda$, where λ is the wavelength. The two components will then add in anti-phase, tending to cancel each other out. Space diversity provides two different receivers and it is hoped that a deep multipath fade will not occur simultaneously at both

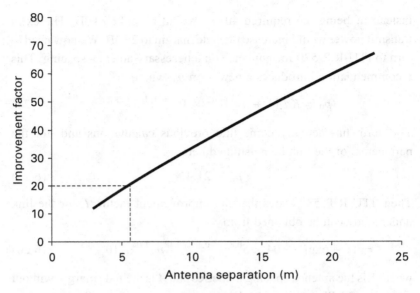

Figure 6.2 The space-diversity improvement factor for the 60-km, 7-GHz link described.

receivers. Remember that different frequency means different wavelength. The hope when using frequency diversity is that the same physical multipath routes will not produce simultaneous deep fades at two separate wavelengths. Note that deploying frequency diversity will entail using twice as much spectrum. The equivalent formula for the improvement factor when adopting frequency diversity is

$$I = \frac{80}{fd}\left(\frac{\Delta f}{f}\right)10^{4/10}.\qquad(6.27)$$

The equation shows that the improvement decreases as distance and frequency increase. The maximum value of $(\Delta f/f)$ covered by this equation is 5% and the improvement predicted for this link even at the maximum frequency separation of 350 MHz is only 4.8. The fact that an extra frequency is required means that frequency diversity is not often used. Frequency diversity is, however, sometimes used in conjunction with space diversity to give additional improvement in availability.

Space diversity and frequency diversity are the two most established forms of diversity on radio links and are supported by well-established methods of estimating the improvement in link performance than can be achieved. Other diversity methods include angle and polarisation diversity, covered briefly below.

6.5.3 Angle diversity

In this case the receiving antennas are co-located but have different principal directions. This is often achieved by adding a second, offset, feed-horn to a parabolic dish antenna. The gain via the offset horn is usually lower than that via the main horn but, if the signal entering the main horn is suffering a deep fade, it can provide a useful back up.

6.5.4 Polarisation diversity

This involves simultaneously transmitting and receiving on two orthogonal polarisations (e.g. horizontal and vertical). The hope is that one polarisation will be less severely affected when the other experiences a deep fade.

6.5.5 Summary of diversity on fixed links

A deep fade can be regarded as a rare event during which the physical structure of the atmosphere has to be exactly right (or wrong) to allow cancelling components to arrive at the receiver. Diversity techniques rely on establishing a second radio path, only very slightly different from the main one. The existence of simultaneous fades on both paths is then thought to be extremely unlikely.

The description of diversity given here is inevitably a simplified version of the content of ITU-R P. 530. For a fuller description, the recommendation itself should be consulted.

One final note: with space-diversity systems the approach used is to transmit through only one antenna and receive at two antennas. If both antennas were used to transmit through, an interference pattern with sharp nulls in it would result.

6.5.6 Diversity in mobile communications

A mobile telephone link has to be balanced. It is of no benefit having a high-powered transmitter at the base station that provides coverage over a long distance in the down-link direction if the power of the mobile terminals is not sufficient to provide the same range on the 'mobile-to-base-station' link (the 'up' link). The mobile terminal is a low-power device. One method deployed to increase the up-link range is space diversity. Horizontally separated antennas a few metres apart can be seen on many mobile-telephone masts. The extra gain on the up link provided by the diversity allows the transmit power (and hence range) of the base station to be increased whilst still maintaining two-way communications.

Because of the multiple reflections that a wave is likely to undergo on its route from the transmitter to the receiver, the polarisation is likely to change. Because of this, it is often deemed necessary to receive on more than one polarisation on the up link. Often, a single antenna unit will in fact consist of two antennas receiving polarisations at ± 45 degrees to the vertical. A combining unit then delivers the best possible signal to the receiver.

6.6 Diffraction fading

The fact that the atmospheric structure affects the way in which radio waves propagate through it has been discussed previously. A common structure is for the refractive index to decrease with height, causing the radio wave to tend to alter its direction so as to follow the curvature of the Earth. This reduces the effect of the Earth's 'bulge' and is accounted for by using a k-factor that magnifies the effective Earth radius. However, the value of the k-factor is not constant. It varies with time. Sometimes the structure of the atmosphere will be such that the refractive index increases with height. This will tend to cause radio waves to bend away from the surface of the Earth, effectively exaggerating the curvature of the Earth (the k-factor is less than unity). If the k-factor is abnormally low, it is possible for the effective earth bulge to obstruct the propagation path. Figure 6.3 shows the effect of varying k-factor on a 30-km link with 20-m antennas at each end.

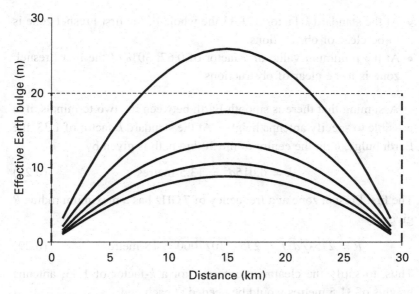

Figure 6.3 The variation in effective Earth bulge for k-factors of 0.67, 1.0, 1.33, 1.67 and 2. The dashed line at a height of 20 metres is shown for guidance.

It can be seen that, at the standard k-factor of 1.33, there is approximately 7 metres of clearance but, at a k-factor of 0.67, the link is obstructed by a similar amount. The likely minimum k-factor experienced on a link is length-dependent. Anomalous atmospheric effects that would lead to very low values of k-factor tend to be very localised. The ITU recommendation P. 530 provides a graph that gives the value of k-factor exceeded for 99.9% of the time as a function of path length. Whereas a link of 20 km in length could have a minimum value of k-factor of less than 0.6, a 100-km link has a minimum value of approximately 0.9. The value given for a link of length 30 km used in the example is approximately 0.67, making the curvatures illustrated in figure 6.3 realistic. When assessing clearance requirements it is important to take into consideration the minimum likely value of k-factor; ITU-R P. 530 recommends that clearance is assessed at both the standard value of k-factor (usually 1.33) and the minimum value that is exceeded for 99.9% of the time.

As an example, consider establishing a 7-GHz microwave link over a distance of 30 km. We are to provide clearance according to two criteria.

- At the standard k-factor of 1.33 the whole of the first Fresnel zone is to be clear of obstructions.
- At the minimum value of k-factor of 0.67, 30% of the first Fresnel zone is to be clear of obstructions.

Assuming that there is smooth Earth between the two terminals, it is possible to specify antenna heights. At the standard k-factor of 1.33, the Earth bulge, h, in the centre of the 30-km path is given by

$$h = 0.015d^2 = 13.5 \text{ metres.} \tag{6.28}$$

The first Fresnel zone at a frequency of 7 GHz has a maximum radius R given by

$$R = 275\sqrt{d/f} = 275\sqrt{30/7000} = 18 \text{ metres.} \tag{6.29}$$

Thus, to satify the clearance criterion for a k-factor of 1.33, antenna heights of 31.5 metres would be needed at each end.

At a k-factor of 0.67, the Earth bulge will double to 27 metres, but we now need to have only 30% of the first Fresnel zone clear; 30% of $18 = 5.4$ metres. Therefore, to satisfy the criterion for clearance at a k-factor of 0.67, antenna heights of 32.4 metres would be needed at each end. The higher of the two heights calculated would be specified. Therefore, in this example, antennas of height 32.4 m would be specified because this would then satisfy both clearance criteria.

6.7 Selective fading in long-distance, high-capacity microwave links

Microwave links occupy a small but finite bandwidth. A multipath fade is a frequency-dependent event. That is, the depth of the fade will vary with frequency (if it did not vary with frequency then frequency diversity would be ineffective). If the variation of the depth of fade within the bandwidth of the link is significant, the spectrum will be distorted. Such distortion will lead to a high bit error ratio, even if the strength of the signal is acceptable. In this situation, simply designing in a fade margin will not help: it is the distortion of the signal that is the problem. As an example let us consider a microwave link at a carrier frequency of 10 GHz with a bandwidth of 100 MHz. Suppose that fading occurs when the radio wave can travel over two different paths with the two contributions arriving in anti-phase. The minimum delay of one path relative to the other that will lead to such a deep fade occurring is half a period of a cycle of the carrier wave. The carrier wave has a period of 100 picoseconds and therefore a relative delay of 50 picoseconds between the two paths would lead to a deep fade. Suppose that the amplitude of one wave was 90% of that of the other. If they are in anti-phase, the resultant would have an amplitude of only 10% of that of the stronger single contribution. This would be the case when there is a fade of 20 dB ($20 \log(0.1) = -20$). The delay of 50 picoseconds results in the signals being exactly in anti-phase only at exactly 10 GHz. Since the bandwidth is 100 MHz, the occupied bandwidth is from 9.95 GHz to 10.05 GHz. At the extremes of the occupied bandwidth, the phase difference would be about 179.1 degrees. The resultant signal amplitude relative to the stronger contribution is then

$$\sqrt{(1 + 0.9\cos 179.1°)^2 + (0.9\sin 179.1°)^2} = 0.101. \qquad (6.30)$$

Now, $20 \log(0.101) = -19.9$ and thus the fade depth at the edge of the band is 19.9 dB, compared with 20 dB at the centre. Thus the spectrum received would be virtually flat and the fading is known as 'flat fading'. This can be mitigated against by ensuring that a suitably large fade margin has been accommodated in the link budget.

This flat fading occurs when the relative delay between the two contributions is very small. The two contributions would be in anti-phase at the centre of the occupied bandwidth if the relative delay were 150 ps, 250 ps, 350 ps etc. Let us consider the situation when the relative delay is 5050 ps (or just over 5 nanoseconds). This represents $50\frac{1}{2}$ periods. At the centre of the band, the phase difference would be $(50 \times 360 + 180) = 18\,180$ degrees, which would still result in a cancellation and a fade of depth 20 dB would again be experienced. At one edge of the occupied bandwidth, the phase difference would be 0.995 of this amount: 18 089 degrees. Of course, 18 000 degrees is the same as

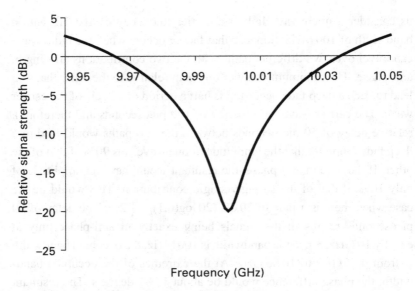

Figure 6.4 Selective fading on a wideband link with a relative delay of 5 ns.

0 degrees phase shift. It is the remaining 89 degrees that is significant:

$$\sqrt{(1 + 0.9\cos 89°)^2 + (0.9\sin 89°)^2} = 1.35. \qquad (6.31)$$

Now, $20\log(1.35) = 2.6$. The signal would be enhanced at the edge of the occupied bandwidth. Figure 6.4 shows the variation in signal strength across the 100-MHz band. This is for a delay of about 5 nanoseconds. For larger delays the problem becomes worse with the signal strength going through peaks and nulls as the frequency varies within the 100-MHz bandwidth. Equally the problem becomes more serious as the bandwidth increases.

Relative delays as large as 5 ns will occur only on longer links (such as a few tens of kilometres). Further, such delays are a problem only if the bandwidth occupied by the radio link is wide (several tens of MHz). But the problem is a serious one. It represents distortion of the radio channel. This will cause the bit error ratio to become unacceptable. Further, such distortion cannot be rectified by simply increasing the power: this will help with flat fading but not with selective fading. The situation is analogous to attempting to understand speech in a large hall that produces a lot of echoes. If the echo is too severe, it will be impossible to understand the speech and increasing the volume will not help the situation. Although building in a fade margin to mitigate against this will be ineffective, implementing a diversity path will yield an improvement.

6.8 Ducting and interference in point-to-area systems

Recommendation P. 1546 considers the effect of ducting by means of curves that show the signal level that is exceeded only for a few per cent of the time. Figure 6.5 shows curves predicting the field strengths exceeded 50% of the time and 1% of the time over a land path at 600 MHz with an antenna transmitting an effective radiated power (ERP) of 1000 watts at a height of 75 metres. At short distances of up to 20 kilometres, the difference between the two curves is very small. However, at distances beyond about 100 kilometres the differences are substantial, indicating that ducting can cause the signal to increase by about 20 dB for 1% of the time on such

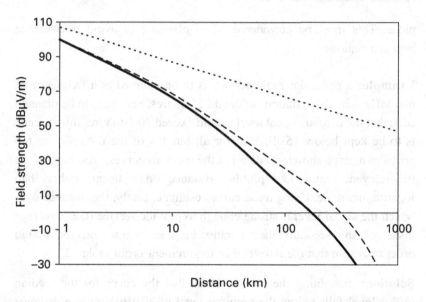

Figure 6.5 A graph showing the predicted median signal strength (solid black line) and strength not exceeded for 1% of the time (dashed line) at 600 MHz (antenna height 75 m).

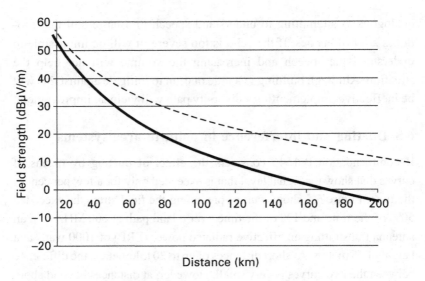

Figure 6.6 A graph showing detail of curves predicting the median signal strength (solid black line) and strength not exceeded for 1% of the time (dashed line) at 600 MHz (antenna height 75 m).

paths. This must be considered when planning to avoid interference between stations.

Example: a particular radio service is to be planned at a frequency of 600 MHz using base stations of height 75 metres. Service is to be planned such that the median signal level should exceed 30 dBμV/m. Interference is to be kept below 15 dBμV/m for all but 1% of the time. Figure 6.6 provides a graph showing a detail of the previous curves, 'zooming in' on the relevant region and plotting distance on a linear, rather than logarithmic, scale. Using these curves estimate, firstly, the distance over which the signal level is strong enough to provide service (the 'coverage distance') and, secondly, the separation between base stations required in order to ensure that the interference requirement is not violated.

Solution: examining the curves shows that the curve for the median (50%) level falls below the required level of 30 dBμV/m at a distance between 50 km and 55 km. Being cautious, a coverage distance of 50 km could be predicted. Now, turning to interference, the curve for levels

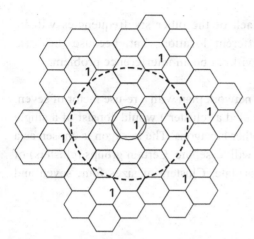

Figure 6.7 Tessellation showing a re-use factor of seven. The solid circle shows the coverage area; dotted circle shows the area over which interference would occur.

exceeded 1% of the time is determined. The maximum level is 15 dBμV/m. To the accuracy possible from examining the graph, the level of the 1% curve reaches 15 dBμV/m at a distance of 170 km. Therefore the minimum separation required between base stations will be 220 km. Each base station would serve an area of 7850 km² but would prevent any users receiving the same frequency from a different base station over an area of 91 000 km². It is said that the base station 'sterilises' an area of 91 000 km². The ratio of the area sterilised to the coverage area is approximately 12 : 1. This ratio is of significance when planning a network providing continuous coverage. The network would require several different frequencies to be allocated in order to provide continuous coverage without unacceptable interference occurring. Interestingly, the required number of frequencies does not necessarily equal 12. One skill of a network planner is to deliver a network that meets coverage and interference requirements whilst requiring the minimum bandwidth. Figure 6.7 shows a possible network plan using only seven different frequencies. Each hexagon is a 'cell' that is served by a base station located at its centre. The solid circle shows the coverage range (diameter 50 km) of the central base station that has been allocated 'frequency 1'. The dotted line shows the area over which interference would occur, making it impossible to re-use that frequency. It can be seen that, although interference occurs within 18 nearby cells, none of them uses 'frequency 1'. Each of the other six frequencies will be interfered with in three different locations but, because they use different frequencies, there will not be an interference problem.

Figure 6.7 shows a classic network plan with a re-use factor of seven. Each group of seven cells (called a 'cluster') would consist of a single hexagon surrounded by six other hexagons. The hexagon is chosen as a shape similar to a circle that will tessellate. Certain groups (clusters) of these hexagons will also tessellate. Clusters of sizes four, seven and twelve are best known.

6.9 Rain attenuation

Rain and other 'hydrometeors' (the collective name for rain, snow and hail) hinder the propagation of radio waves. Rain absorption is very frequency-dependent. Below about 5 GHz it can largely be ignored. It does not therefore feature largely in planning methods for services such as broadcasting and mobile telecommunications. For fixed links at frequencies above about 20 GHz, however, it is often the most significant factor in limiting the maximum possible path length. Of course, the effect of rain depends on the rainfall rate. The most significant statistic is the peak rainfall rate (that which is exceeded for only 0.01% of the time) in a region, rather than the total annual rainfall. This is very geographically dependent. In dry regions, this rate may be only a few millimetres per hour. Regions that experience monsoon rains will have peak rainfall rates as high as 150 millimetres per hour. Temperate regions have peak rainfall rates of around 40 millimetres per hour. The ITU publishes detailed methods of predicting the rainfall attenuation in dB. A fixed link of path length 10 km operating at a frequency of 20 GHz will experience additional path loss of about 1 dB for every 1 mm/h of rainfall. Horizontally polarised waves are found to be subject to higher loss than vertically polarised waves (with a typical difference being 20% of the loss in dB). At a rainfall rate of 40 mm/h, loss on a 10-km path would be negligible below 5 GHz, about 10 dB at 10 GHz, about 40 dB at 20 GHz and as much as 160 dB at 80 GHz.

A radio wave will be both absorbed and reflected by rain drops as it passes through a rain storm. Because of the nature of rain, normal diversity techniques are ineffective at mitigating against rain fades.

When designing a radio link, it is normal to design it for a certain percentage unavailability, usually 0.01% of the time. A crucial input parameter to our system-design process is therefore the rain rate exceeded for 0.01% of the time. Note that this does not necessarily relate directly to annual rainfall. Areas with fairly constant low rainfall rates present a lower peak loss than do areas with low annual rainfall but that suffer short intense bursts of rain.

Conversion from peak rain rate to the loss on a given link is a multi-step process.

1. Determine the peak rain rate (either from local data or using ITU-R P. 837).
2. For the frequency and polarisation being used, determine the loss per kilometre (using ITU-R P. 838).
3. For the path length and rain rate determine the effective path length.
4. Multiply the effective path length by the loss per kilometre to determine the peak link loss.

The loss per kilometre is very dependent on frequency and is somewhat dependent upon polarisation, with horizontally polarised signals being affected more than vertically polarised signals. Figure 6.8 shows a graph of loss per kilometre against frequency for both vertical and horizontal polarisation for a peak rain rate of 40 mm/h (as is typical for northern Europe). Figure 6.9 shows how the loss increases with rain rate at a frequency of 20 GHz. The relationship can be seen to be fairly linear.

The additional path loss in dB/km is given the term γ. The peak total loss along a path can be estimated by considering what is known as the 'effective path length'. The principle is that, although the rain may be falling at the peak rate at one point along the path, it is unlikely to be falling at the peak rate at all points on the path, particularly if the path is long. The effective path length, d_{eff}, is calculated (from ITU-R P. 530) as

$$d_{\text{eff}} = \frac{d}{1 + d/d_0},\tag{6.32}$$

where $d_0 = 35\mathrm{e}^{-0.015R_{0.01}}$, $R_{0.01}$ being the rain rate exceeded for 0.01% of the time. The total additional path loss due to rain is then determined from the equation

$$\text{loss} = \gamma d_{\text{eff}} \text{ dB.} \qquad (6.33)$$

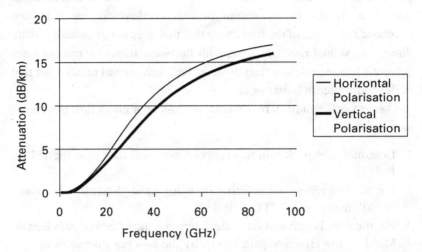

Figure 6.8 The variation of additional path loss with frequency for a rain rate of 40 mm/h.

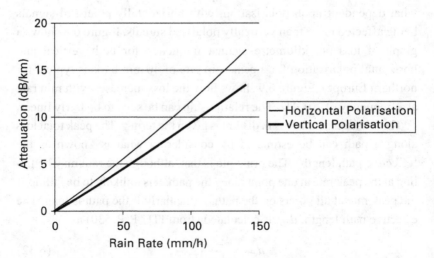

Figure 6.9 The variation of additional path loss with rain rate at a frequency of 20 GHz.

Example: the peak, 0.01% rain rate for a location in south-east England is 28 mm/h. It is required to establish a 16-km link utilising the 20-GHz band. Vertical polarisation is to be adopted. Determine the peak rain attenuation that will be suffered for 0.01% of the time.

Solution: from the graph of figure 6.8, the loss at a rain rate of 28 mm/h is approximately 3 dB/km for vertical polarisation. The intermediate parameter

$$d_0 = 35e^{-0.015 \times 28} = 23 \text{ km} \tag{6.34}$$

and therefore

$$d_{\text{eff}} = \frac{16}{1 + 16/23} = 9.4 \text{ km} \tag{6.35}$$

and the rain attenuation is given by

$$\text{rain attenuation} = 3 \times 9.4 = 28.2 \text{ dB}. \tag{6.36}$$

6.10 Atmospheric absorption

The atmosphere is not a lossless medium. The molecules present in it (particularly water vapour and oxygen) have resonance frequencies at which high losses occur. Figure 6.10 shows the variation of loss with frequency for an atmosphere of typical temperature, pressure and water-vapour content (such an atmosphere is referred to as the 'mean global reference atmosphere'). Generally speaking, atmospheric loss is negligible (less than 0.1 dB/km) at frequencies below 10 GHz. Above

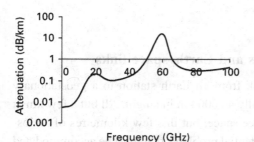

Figure 6.10 A graph showing the variation of attenuation per kilometre with frequency for a standard global atmosphere.

10 GHz, the loss rises steadily from about 0.1 dB/km at 10 GHz to 0.2 dB/km at 20 GHz. Above 20 GHz it does not vary significantly until a frequency of about 58 GHz is reached. At this frequency, atmospheric loss increases sharply (at a resonance frequency of the oxygen molecule), until at about 60 GHz the loss is approximately 17 dB/km. The highly absorbing band of frequencies around 60 GHz is known as the absorption band. Above 60 GHz, absorption peaks are observed together with lower-loss 'windows' in between them. Full details of atmospheric absorption are given in ITU-R P. 676. Because the loss is partly due to absorption by water vapour and partly due to absorption by oxygen, the loss does depend upon the humidity. The loss at 20 GHz is mostly due to water vapour. The global reference atmosphere will exhibit losses of about 0.2 dB/km at 20 GHz whereas dry air will produce a loss of only 0.1 dB/km at that frequency. Conversely, the large absorption peak at around 60 GHz is due to absorption by oxygen and dry and humid air will each exhibit a similar absorption of about 17 dB/km. Notice that the value of attenuation is plotted on a logarithmic scale, thus tending to compress the extreme values of loss. The values quoted here are the losses that would be experienced at about sea level. At higher altitudes, there is less atmosphere and the loss will be lower. However, the effect is not dramatic. At altitudes of 5000 metres, the loss will be 80% (in dB) of the loss at sea level.

Atmospheric absorption is another loss that must be considered when designing a link. It is an 'ever-present' loss and must be added to any other losses such as those predicted to occur during rain. The high losses experienced in the absorption band render it useful for secret, short-distance communications and for secret communications between high-flying aircraft or satellites.

6.11 Atmospheric effects and Earth–space links

Consider, for example, a link from an Earth station to a geostationary satellite; such links are typically 40 000 km in length. All but a few km at the Earth-station end is in free space, but this few kilometres of atmosphere is sufficient to cause potential problems that must be accommodated

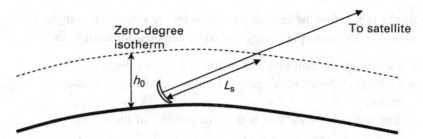

Figure 6.11 A path geometry for determining rain-induced losses on an Earth–space link.

when designing such a link. The most commonly used frequencies for links to geostationary satellites are

- uplink 6 GHz, downlink 4 GHz;
- uplink 14 GHz, downlink 11 GHz;
- uplink 30 GHz, downlink 20 GHz.

Since atmospheric absorption at these frequencies is at most a few tenths of a dB per km, absorption as such does not have a great impact on the link budget. Absorption by rain can, however, be significant, particularly at the higher frequencies listed. Figure 6.11 shows the geometry of a link between an Earth station and a geostationary satellite. Of interest is the height of the 'zero-degree isotherm', h_0, at a particular location. Above the zero-degree isotherm, liquid water does not exist and the atmosphere is dry and, interestingly, hydrometeor absorption loss is very nearly zero. The average height of the zero-degree isotherm varies from about 1 km to 5 km depending on location on the surface of the Earth. Recommendation ITU-R P. 839 contains a map of the world showing heights of the zero-degree isotherm and, further, refers users to a database providing more accurate information. The distance that the radio wave has to travel below the zero-degree isotherm affects the loss because water exists in liquid form in this region. Since the radio wave travels at oblique incidence through this region it is known as a 'slant path'. The equations used in recommendation P. 618 to estimate the additional path loss on Earth–space links due to rain below the zero-degree isotherm are somewhat complicated. However, the principle is very similar to that used for fixed links.

- Determine the rain rate and establish a loss in dB/km.
- Determine the effective path length by calculating a reduction factor.
- The additional path loss is then the product of these two.

As an example, let us consider an Earth station that looks at a geo-stationary satellite with an elevation angle of 20 degrees (see figure 6.11). The peak rain rate is 40 mm/h and the zero-degree isotherm has a height of 2 km. At this rain rate the effective length of the slant path is 80% of the actual length. We wish to calculate additional loss on the downlink at frequencies of 4 GHz, 11 GHz and 20 GHz for vertical polarisation.

Firstly, the slant path length can be calculated from

$$L_S = 2/\sin 20° = 5.8. \tag{6.37}$$

The effective path length used to predict additional loss due to rain is 80% of this value: 4.6 km.

Now, ITU-R P. 838 is used to determine a value of attenuation per kilometre at the frequencies and polarisation described. In order to determine accurate values of attenuation, the recommendation itself should be referred to. However, the graph given in figure 7.1 in the next chapter gives approximate values of the loss for a rain rate of 40 mm/h. At 4 GHz, the loss is negligible. At 11 GHz, the loss is approximately 1 dB/km, giving an estimated peak additional loss due to rain of about 5 dB. At 20 GHz, the loss is approximately 3.5 dB/km, giving a total loss of approximately 16 dB.

The variation of loss with frequency is clearly demonstrated. Space diversity can mitigate against this loss, but only if the separation between Earth-based antennas is large enough for there to be a reasonable chance of the rain rate being considerably lower in one location when it is raining heavily in the other. Separations of at least several kilometres are needed to give a significant diversity improvement. Then, combining the two signals becomes a challenge.

6.12 Noise and temperature effects

When it is considered to be operating as a receiver, an antenna is thought of as a device that gathers signal power. It is also, however, a device that gathers noise power. Just as it gathers a signal dependent on the direction in which it is pointing, so it gathers noise power dependent on its direction. Every molecule radiates radio noise. The amount of radio noise any molecule radiates depends upon its temperature. The noise gathered by an antenna therefore depends on the temperature of whatever it is looking at. This makes analysis quite simple for terrestrial links from one point of the Earth's surface to another: everything that the antenna can see is at approximately the same temperature. When calculating noise, temperatures are recorded in degrees Kelvin (or 'kelvins'). Note that, to obtain the temperature in kelvins, you simply add 273 to the temperature in centigrade. The noise power gathered by an antenna is then found to be kTB watts where k is Boltzmann's constant (1.38×10^{-23} joules per kelvin), T is the temperature in kelvins and B is the system bandwidth in hertz. For terrestrial systems, a standard temperature of 290 kelvins (17 °C) is used to estimate the noise power entering the antenna. For a bandwidth of 10 MHz, the noise power would then be $1.38 \times 10^{-23} \times 290 \times 10^7 = 4.0 \times 10^{-14}$ watts or -104 dBm. This is known as thermal noise. It is this thermal noise that determines the fundamental limit on all telecommunications systems.

All receivers will add internally generated noise to that gathered by the antenna. The receiver is characterised by its effective noise temperature, T_e. The total noise at the input can then be regarded as $k(T + T_e) B$ watts. Calculations can then be carried out as if the receiver were perfect.

Determining the noise temperature for satellite Earth stations, or radio telescopes, is more complicated. The antenna is looking at deep space, whose temperature is as low as perhaps 4 kelvins, but there will be some atmosphere at a much higher temperature in the way. The effect of the intervening atmosphere on the noise temperature can be determined if its loss is known. Clouds etc. therefore have a double effect on such systems: they attenuate the wanted signal and add to the noise. Wherever possible, radio telescopes are situated at high altitudes to minimise the atmospheric effects.

6.12.1 Noise calculations on Earth–space systems

When a signal from a satellite is received by an Earth station, the ratio of the wanted signal power to any unwanted power is a crucial parameter in determining the quality of any received signal. Thermal noise is a significant contributor to the unwanted power. The Earth station's main beam is directed into space. Deep space has a temperature of only perhaps 4 K. The minimum possible noise power that it will gather in a bandwidth of B Hz is therefore kTB with T equal to 4. However, as well as gathering noise power through the main beam, some power will be gathered by side lobes that are directed at the Earth, which will have a much higher temperature. The total noise power is an aggregate of all these losses. Whilst the effect of side lobes will increase the total noise power gathered, the overall noise temperature is substantially increased by losses in the system. These can take the form of resistive losses in the antenna and feeder system and atmospheric losses, especially higher losses produced by clouds and rain in the atmosphere.

To understand the effect of resistive losses on noise temperature, it is necessary to conduct a thought experiment. Let us compare two matched systems with all input components at the standard temperature of 290 K. The first system has a lossless input and the only noise at the input is therefore kTB watts. Now, the term noise temperature is a useful way of predicting the signal-to-noise ratio at the output of a receiver. It has no physical meaning as such and certainly cannot be taken as an indication of the actual temperature of any element of a communications link. The second system has an attenuator at the input. The system noise will be exactly the same but the signal will be attenuated and thus the signal-to-noise ratio will be lower. This effect can be described by attributing a noise temperature to the attenuator. Suppose that the attenuator produced 2 dB of loss. This will reduce the signal-to-noise ratio by 2 dB. The effect is equivalent to the total noise power being increased by 2 dB. Since 2 dB is a ratio of 1.68, the total effective noise power is $1.68kTB$ watts. Thus the effect of the attenuator is to add $0.68kTB$ watts to the total noise power. Thus the effective noise temperature of the attenuator equals $0.68T$, where T is 290 K in this case. The general formula is that the effective noise temperature of an attenuator, T_e, is given by

$$T_e = (L - 1)T, \qquad (6.38)$$

where L is the loss of the attenuator as a ratio (rather than in dB) and T is the temperature of the attenuator in kelvins.

If the loss is at the input of a receiver with a very low noise temperature, the effect of a small amount of loss can be dramatic. Amplifiers of very low noise temperature can be produced by cooling the components to a very low temperature, perhaps using liquid helium. Suppose that an amplifier had a noise temperature of only 40 K. Now consider the effect on the signal-to-noise ratio if resistive losses at the input amounted to 1 dB and the components producing these losses were at room temperature, 290 K. Since 1 dB is a ratio of 1.25, the noise temperature of the lossy component will be $0.25 \times 290 = 73$ K. Thus the noise temperature of the system will have increased from 40 K to 113 K because of these losses. This is an increase by a ratio of 2.8, or 4.5 dB. Thus the insertion of only 1 dB of loss has reduced the signal-to-noise ratio by 4.5 dB. In reality, this figure of 4.5 dB represents the fact that the attenuator will do two things: it will attenuate the signal and it will actually generate extra noise by the motion of its molecules that are at a higher temperature than the rest of the system. The reduction of 4.5 dB in signal-to-noise ratio incorporates both of these effects.

Let us consider now a highly sensitive Earth station whose main beam is pointed at deep space with a noise temperature of only 4 K, but with side lobes pointed at the Earth. The overall noise gathered by the antenna corresponds to a temperature of 30 K. However, resistive losses in the reflective parabolic dish and the feed system amount to 0.5 dB. The dish and feed are at a temperature of 290 K. Since 0.5 dB is a ratio of 1.12, the noise temperature caused by the losses is $0.12 \times 290 = 35$ K. Thus the total effective noise temperature is now 65 K, an increase by a ratio of 2.18, or 3.4 dB.

Suppose now that this level of resistive loss is accepted and the antenna system is regarded as having a noise temperature of 65 K. Let us suppose that a rain storm introduces 4 dB of loss. Loss produced by a rain storm is a resistive loss (the water molecules will generate thermal noise). Since 4 dB equals 2.5 as a ratio, the effective noise temperature is $1.5 \times 290 = 435$ K. The total effective noise temperature is $435 + 65 = 500$ K. The increase from 65 K to 500 K constitutes an increase by a factor of 7.7, or 8.9 dB. Thus rain attenuation of 4 dB would make the signal-to-noise ratio 8.9 dB worse.

When dealing with extremely sensitive receivers such as those used for radio astronomy, gaseous attenuation can become a significant factor in determining the performance of the receiving system. For this reason it is often seen as desirable for radio telescopes to be situated at high altitudes in order to minimise the amount of atmosphere between them and the celestial object under investigation.

6.13 Summary

The structure of the atmosphere can have a significant effect on the nature of the received signal in a radio link. The two major effects are referred to as 'multipath propagation' and 'ducting'. Under conditions where multipath propagation occurs the received signal power can drop very suddenly by tens of dB. The probability of such fading of the signal occurring can be estimated using parameters such as path length, frequency, path inclination and a further climatic factor that depends on the location on the Earth's surface. Ducting produces enhancements of signal strength at large distances (the effects are most noticeable beyond the horizon, where the signal strength is normally very low). Ducting is responsible for many incidences of interference over long-distance paths.

The effect of multipath fading can be mitigated against by deploying a diversity system. One common example, space diversity, entails using a second receiving antenna in the expectation that a deep fade will not simultaneously affect both antennas. Diversity-based systems are widely used on long-distance, high-capacity, microwave links because these suffer from selective fading whereby the spectrum of the received signal becomes distorted.

Another significant atmospheric effect is that of attenuation due to rain. Below about 10 GHz, rain fading is not very significant, but, at higher microwave frequencies, it becomes the major factor limiting path length, particularly in areas that experience high levels of rainfall. Additionally, the atmosphere itself is not perfectly lossless. Again the effect increases with frequency, with particularly high levels of loss occurring at about 60 GHz.

Thermal noise presents a limitation to all communication systems. In radio systems thermal noise is gathered alongside the wanted signal by the receiving antenna. The amount of noise depends on where the antenna is pointed and the loss of the atmosphere. Thermal noise levels are lower when the receiving antenna is pointed into space, but atmospheric and system losses then degrade the noise performance of the system.

7 System design and interference management

The test of the usefulness of the knowledge gained can best be determined by undertaking some practical exercises that give an insight into problems encountered by radio system designers in the 'real world'. Firstly, the value of propagation studies in helping to identify the most appropriate frequency for various services is discussed. The system design of microwave links at 10 GHz (at which multipath fading will dominate) and at 23 GHz (at which rain fading will dominate) is explained in some detail. The fact that many thousands of microwave links will be required in an industrialised country leads to a need for interference management, so this topic is introduced. Next, attention is turned to the design of broadcasting systems with a view to obtaining maximum coverage whilst investigating methods of limiting interference at great distances. Additionally, an example of designing a link to a geostationary satellite is presented. Finally, special methods needed for providing and predicting the signal strength for in-building systems are presented.

7.1 Determining the most appropriate frequencies for specific services

We have seen that the frequency of operation affects the way in which a radio wave is affected both by obstacles and by the atmosphere and rain. At first sight, it seems that the lower the frequency, the better. Obstacles cause lower levels of diffraction loss at longer wavelengths. Also, the effect of rain and atmospheric absorption is almost negligible below about 5 GHz. Further, the penetration of materials such as concrete is better at lower frequencies. It is, however, important to consider two additional facts: antennas tend to be larger at lower frequencies and, further, users of the radio spectrum are expecting to communicate at ever-increasing bit rates, demanding large amounts of bandwidth.

Microwave antennas, which are based on the parabolic dish reflector, collimate the energy into a narrower beam as the frequency increases. For a given size of antenna, the link loss decreases as frequency increases (by 6 dB for every doubling of frequency). This suggests that, for point-to-point services, the higher the frequency the better. If it were not for the effect of cloud and rain (and the fact that the development of higher-frequency amplifiers and receivers presents technical challenges), that statement would be true. It is true that radio links that avoid obstacles are better suited to higher frequencies whereas those where obstacles may be expected are better suited to lower frequencies. Further, if the terminals at each end of the link are constrained not so much by size but by antenna radiation-pattern requirements (such as with mobile radio systems) then the link loss will be lower at lower frequencies. This is offset by the need for high throughput and hence high bandwidth.

One inescapable fact is that the bandwidth available increases with increasing frequency. A typical GSM mobile network will need approximately 20 MHz of bandwidth in order to provide continuous coverage and satisfy user requirements for connection on demand. If the frequency used is too low, it will be impossible to offer the necessary bandwidth (especially when four or more operators have similar bandwidth requests). If the frequency is too high, the effect of obstacles and link loss to terminals with omni-directional antennas will be such as to make it difficult to provide continuous coverage across a region. This leads to something of an optimum range of frequencies for mobile communications: roughly in the 300–3000-MHz range. Above 3000 MHz point-to-point services (including from Earth stations to satellites) dominate use of the spectrum. Long-distance broadcasting and international terrestrial communication can use frequencies from about 30 kHz up to 30 MHz. The bandwidth allocated is small, allowing for moderate-quality audio, or low bit rates, only. Broadcasting above 30 MHz will cover a shorter range and can be allocated a larger bandwidth to provide higher-quality audio coverage. Because of the large bandwidth requirements for television transmission (just less than 10 MHz), a higher frequency is allocated, usually within the 400–900-MHz range, making it rival mobile services for frequency allocations.

7.2 System design

It is common to hear about the spectrum being referred to as a 'finite resource'. Strictly speaking, it isn't: there is no 'highest frequency'. However, we have seen that as frequency increases losses due to rain increase and the ability to diffract around obstacles decreases. Thus spectrum can be argued to become less desirable as frequency increases. On the other hand, it is easier to control the electromagnetic wave using reasonably sized antennas as frequency increases. Also, most importantly, the bandwidth available increases with frequency (you cannot allocate 200 MHz of bandwidth at a frequency of 50 MHz). For this reason, there is a region of the spectrum that is seen as most desirable from the viewpoint of being suitable for high-capacity mobile or broadcast purposes using antennas of a manageable size. Since spectrum for mobile services, in particular, is seen as particularly desirable, this leads to a particular region of spectrum (typically between about 300 MHz and 3000 MHz) being very highly valued. Television broadcasting, mobile telephony and data services and long-distance microwave links are all located in this frequency range. Note that these are all terrestrial (land-based) systems. One of the reasons why frequencies below 3000 MHz are desirable for long-distance point-to-point links is that such links are less susceptible to atmospheric and rain effects. However, the lower the frequency, the larger the antenna needed. If the atmosphere did not play such a big part, the idea of the most desirable frequency would change. An example of this is satellite links.

In a link to a geostationary satellite, the path length is approximately 37 000 km, but there is very little atmosphere in the way. Further, in order to be able to position many satellites using the same frequency band in geostationary orbits without interference, it is necessary for the Earth-station antenna to have a narrow beamwidth. For a given size of antenna, the higher the frequency the narrower the beamwidth. When the desire for high bandwidth is added to the list of considerations, the most desirable frequency for communication with geostationary satellites is higher than that for terrestrial systems. Notice that we are considering geostationary satellites that are in orbit at a height of 36 000 km above the equator. Communication with 'low-Earth-orbit' (LEO) satellites for mobile communication or navigation purposes uses lower frequencies (again usually in the 300–3000-MHz range).

An additional feature regarding Earth–space communications is that there is a wide gap between the frequencies used for the uplink and the downlink. This is because there is a big difference between the transmit and receive powers. The most widely used frequencies for communication with geostationary satellites are 6 GHz on the uplink paired with 4 GHz on the downlink and 14 GHz on the uplink paired with 11 GHz on the downlink. There is little rain fading (although it is not negligible) on an Earth–satellite path using 4 GHz or 6 GHz and, therefore, the fade margin can be small. Nevertheless, because of the large distance, the size of antennas required is large. On a terrestrial microwave link, the antennas at each end of the link are likely to be of the same size. On an Earth–satellite link, the antenna at the Earth station will often be much larger than the antenna on the satellite. Earth-station antennas can be as large as 30 metres in diameter when operating in the 4 GHz or 6 GHz band.

For short-distance point-to-point links, a high level of rain fading can be tolerated and the small antenna sizes and high bandwidth associated with the higher microwave frequencies are attractive. For this reason, frequencies up to 40 GHz are widely used for such links, with even higher frequencies beginning to be exploited.

7.2.1 Terrestrial point-to-point system design

When designing a microwave link, an appropriate fade margin must be determined, depending on the frequency and path length. This has to be considered when designing the link budget. This budget will determine the transmit power and size of antennas required to ensure that the received power remains adequate even when a fade occurs. Once this has been determined, a channel allocation will be required. A channel must be selected after interference analyses have demonstrated that the new link will neither suffer harmful interference from, nor cause interference to, existing microwave links or satellite Earth stations.

Example 1: link design at 7 GHz

We can assume that we are required to design a system to carry a channel of a particular capacity over a certain distance. Suppose that a 16-Mbit/s link is required over a distance of 40 km. We will need to specify the antenna size and height together with the transmit power needed to provide a specified availability (typically 99.99%). We will assume that the rain rate exceeded for 0.01% of the time is 40 mm/h. Consideration of the graph shown in figure 7.1 shows that the attenuation expected increases greatly as the frequency exceeds 10 GHz. Therefore, for a long-distance link such as this one, a frequency below 10 GHz will probably be selected (say 7 GHz). At this frequency, the major source of fading will be multipath fading. We need to use ITU-R P. 530 to help us establish a suitable fade margin.

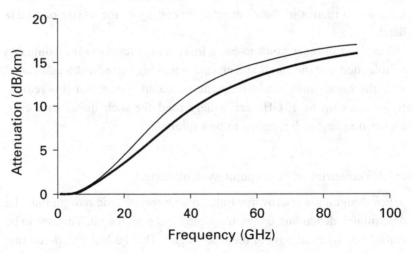

Figure 7.1 The variation of rain attenuation with frequency for a rain rate of 40 mm/h for vertical polarisation (solid line) and horizontal polarisation (thin line).

One factor that must be considered is the radio climatic factor for the region involved. This gives us the crucial parameter, in recommendation P. 530, K. Information that leads us to this value can be found by examining the geographical data given in ITU-R P. 453. This gives information on the rate of variation of the refractive index of the atmosphere over the first 100 metres of altitude, dN_1. This has been found to be a reliable indicator of the probability of multipath fading. Let us assume that the area we are concerned with has the typical value of -200. Other parameters that affect the probability of fading include the frequency (selected to be 7 GHz), the distance (40 km) and the antenna heights. The height above sea level should be expected to affect the probability of fading because, the less atmosphere there is, the lower the chances of atmospheric effects. Additionally, since the atmosphere tends to form itself into horizontal layers, multipath effects are less likely if the antennas are of different heights and the radio wave does not propagate horizontally along the path.

In determining the necessary height above sea level of the antennas, we need to know the height of the terrain under the path. If this is fairly level at a constant height, the antennas have to be high enough to accommodate Earth bulge and, further, must provide sufficient clearance. Designers adopt various clearance criteria, but a fairly standard criterion when designing links over fairly level terrain is that the first Fresnel zone must be clear when Earth curvature at a k-factor of 1.33 has been considered.

The Fresnel zone is an imaginary ellipsoid such that a route from one terminal to the receiver via the surface of the zone is exactly half a wavelength longer than the direct route between the terminals (this is, in fact, the definition of the 'first Fresnel zone'; if the path-length difference were exactly one wavelength, the ellipsoid would represent the 'second Fresnel zone'). The radius of the first Fresnel zone is greatest in the centre of the path and equals $\sqrt{\lambda d}/2$ in 'self-consistent' units (that is, if λ and d are in metres then the radius will be in metres). At 7 GHz, λ is 0.043 metres. Thus, in this case, the radius of the first Fresnel zone will be $\sqrt{0.043 \times 40\,000}/2 = 20.7$ metres.

We have previously derived the Earth bulge over a path of length d for a k-factor of 1.33 to be $0.015d^2$. Thus, for a path of 40 km, an Earth bulge of 24 metres should be considered.

Thus, if the ground in between the two antennas is of approximately the same height, the antennas should each be approximately 45 metres above ground height. Fading statistics depend upon the height above sea level rather than the height above ground level, so we have to assume a certain ground height. Let us assume that the ground is low-lying, at 5 metres above sea level. Thus the antennas should each be 50 metres above sea level.

We now have sufficient parameters to estimate the required fade margin for an availability of 99.99%. We shall use the equations suitable for quick planning applications given in ITU-R P. 530. This gives the probability p_w as a percentage of the occurrence of a fade of depth A:

$$p_w = Kd^{3.0}\left(1 + |\varepsilon_p|\right)^{-1.2} \times 10^{-0.033f - 0.001h_L - A/10}, \qquad (7.1)$$

where h_L is the height of the lower antenna above sea level, in metres. The term ε_p is an indication of the slope of the path in milliradians and is given by

$$|\varepsilon_p| = |h_r - h_e|/d, \qquad (7.2)$$

where h_r and h_e are the heights of the two antennas in metres and d is the path length in kilometres.

The first part of the process is to determine the appropriate value of K to use for the location of this particular link. To determine K we need to use another equation in P. 530. Again, a choice of equation is available and we shall use the method put forward for a quick calculation. (The more exact method requires a determination of the standard deviation of terrain heights along the path – not an easy job without a computer programme. The general rule is that, the smoother the path, the more likely the occurrence of multipath fading.) The 'quick-calculation' equation is

$$K = 10^{-4.2 - 0.0029 dN_1}, \qquad (7.3)$$

where dN_1 is a parameter that describes how quickly the refractive index varies with height. Taking the value of dN_1 to be -200, as stated previously, leads to a value for K of $10^{-4.2 - (0.0029 \times -200)} = 2.40 \times 10^{-4}$. We now have all the information we need to find the probability of a fade of depth A. On substituting into the equation we have

$$p_w = 2.40 \times 10^{-4} \times 40^3 \times 10^{-(0.033 \times 7)-(0.001 \times 50)-A/10}$$
$$= 8.03 \times 10^{-A/10}. \tag{7.4}$$

Hence,

$$A = -10 \log_{10}\left(\frac{p_w}{8.03}\right). \tag{7.5}$$

Thus, if we target a percentage outage of 0.01%, we can determine the required fade margin as

$$A = -10 \log_{10}\left(\frac{0.01}{8.03}\right) = 29.0 \text{ dB}. \tag{7.6}$$

Let us pause to reflect a while on what we have achieved thus far. Our 40-km link will require antennas approximately 45 metres above ground level at each end, assuming fairly level ground, and we need to accommodate a fade margin for multipath fading of about 29 dB. Since rain attenuation is very low at this frequency (7 GHz), we do not need to consider it. Note that you do not have to add a rain fade margin to a multipath fade margin because you do not get multipath fading during a rainstorm: it is therefore common to take the higher of the two fade margins. Strictly speaking, rain fading on this link may increase the unavailability to very slightly more than 0.01%, but the amount by which it does this will be negligible. Thus, we now simply have to design a free-space system but add in the fade margin of 29 dB.

A necessary step is to determine the minimum required signal level at the receiver. In practice, you would look in the equipment manual for this information. However, it is beneficial to have some idea of the level that is likely to be required. You will hear a lot of mention of signal-to-noise ratios, which vary from system to system. A parameter that varies by much less in digital systems is the ratio of the 'energy per bit' to the 'noise spectral density' E_b/N_0. In terrestrial (land-based) systems the noise spectral density has a value of kTF watts per hertz, where k is Boltzmann's constant of 1.38×10^{-23} joules per kelvin, T is the noise temperature (typically assumed to be a standard temperature of 290, known as T_0, kelvins) and F is the noise figure of the receiving system (a ratio). As a rule of thumb a ratio of 'energy per bit' to kT of about $100:1$ (or 20 dB) is sufficient to provide a data channel of sufficiently low bit error ratio. Now

$$kT = 1.38 \times 10^{-23} \times 290 = 4.00 \times 10^{-21} \text{ joules.} \qquad (7.7)$$

Therefore a radio receiver will require an energy of approximately 4.00×10^{-19} joules for every bit of data that is transferred (although you will find variations in practice, this is not a bad estimate). We have said that the radio link will carry data at 16 megabits/second. Thus the required receive power is 6.4×10^{-12} watts, which equals -112 dBW or -82 dBm. Thus a 16-Mbit/s terrestrial radio link will require a receive power of about -82 dBm. Since we are going to design our link to accommodate a fade of up to 29 dB, the target unfaded receive level must be -53 dBm.

Now, the free-space loss for a 40-km link at 7 GHz is

$$92.4 + 20 \log(40) + 20 \log(7) = 141.3 \text{ dB.} \qquad (7.8)$$

The major parameters that will complete the link design are the transmitter power and antenna size. To establish an achievable transmitter power at reasonable cost, we will have to consult the product catalogues, but 1 watt (30 dBm) is a reasonable assumption at 7 GHz (the power achievable tends to reduce with increasing frequency). Thus, in order to establish an unfaded signal level of -53 dBm, antenna gains must be chosen so that

$$30 + G_t + G_r - 141.3 = -53. \qquad (7.9)$$

If we decide on equal gains for the transmit and receive antennas, a gain of 30 dBi for each antenna would be suitable. We will now estimate how large these antennas would have to be to achieve such a gain:

$$\text{antenna gain (dBi)} \approx 18 + 20 \ \log_{10} f \text{(GHz)} + 20 \ \log_{10} D \text{ (metres)}$$
$$30 = 18 + 20 \ \log_{10} 7 + 20 \ \log_{10} D$$
$$D = \frac{10^{12/20}}{7} = 0.57 \text{ m.} \qquad (7.10)$$

Thus an antenna size of 60 cm would be appropriate. This antenna would have a beamwidth of approximately 5 degrees. The system parameters are summarised below:

- antenna size 60 cm diameter,
- antenna gain 30 dBi,
- transmitter power 30 dBm,
- antenna height above ground level 45 m,
- path length 40 km,
- availability 99.99%.

Interference prevention: the need for licensing and assignment

The calculation performed made has been carried out 'in isolation'. That is, the effect of any other users of the radio spectrum was ignored. In the real world due account must be given to the fact that other transmitters and receivers will be operating, often nearby.

In this example, a minimum receive level of -82 dBm was used. If the receive signal level drops below this then the bit error ratio will become excessive. The level -82 dBm is often referred to as the 'threshold' of the receiver. It is based on a required value of E_b/N_0, which in turn is based on the thermal noise level and the required signal-to-noise ratio for the modulation scheme employed. For example, if 'quadrature phase-shift keying' (QPSK) were used as the modulation scheme, a bandwidth of approximately 14 MHz would be required. For a terrestrial system at 7 GHz, an overall noise temperature of about 1000 kelvins would be typical. Thus the noise power would be

$$kTB = 1.38 \times 10^{-23} \times 1000 \times 14 \times 10^6 = 19.4 \times 10^{-14} \text{ watts}$$
$$= -97 \text{ dBm.} \tag{7.11}$$

Thus a signal-to-noise ratio of 15 dB would be achieved. However, if the transmitted power were set to exactly 30 dBm, 99.99% availability would be achieved only if there were no interference. But if another link is allowed to use the same frequency, there will be some interference. This will add to the thermal noise and 'degrade' the threshold. Somewhat confusingly, if a receiver suffers threshold degradation, the threshold level increases. Suppose that interference equal to the thermal noise level of −97 dBm was experienced. The total of noise plus interference would be −94 dBm; 3 dB higher than the interference-free level. In this situation the threshold level would similarly increase by 3 dB, to −79 dBm (a 'threshold degradation' of 3 dB). To compensate for this threshold degradation, the transmit power would have to be increased by 3 dB. But this would lead to extra interference being caused elsewhere. It is the job of the regulator to manage the interference such that a near-optimum situation is reached with a maximum number of users able to use a high-quality communication channel using a minimum amount of spectrum. The two extreme positions that a regulator could adopt are as follows.

- No interference is to be tolerated: all users are to have exclusive access to the spectrum allocated to them.
- Allow users to transmit at as high a power as necessary to overcome any interference.

An optimum position almost certainly lies between these two extremes. One possibility that could be adopted is to include an interference allowance in the link budget and then ensure that the interference experienced in practice does not exceed this allowance. The choice of an appropriate allowance is somewhat arbitrary, but it should be borne in mind that a very low allowance will result in spectrum being lightly used whereas a high allowance (that must form part of the link budget) will result in the maximum path length possible for a particular configuration being reduced. Suppose that we adopt a policy of incorporating an interference allowance that would cause a threshold degradation of 2 dB. In this situation, we would increase the transmit power to 32 dBm. Checks would have to be carried out to ensure that any existing transmitters would not cause threshold degradation of more than 2 dB, such that the total of noise and interference would not exceed −95 dBm. Now the −97 dBm noise level equals 2×10^{-13} watts; −95 dBm equals 3.2×10^{-13} watts. Thus the total interference power must be limited to 1.2×10^{-13} watts. Checks would have to be carried out on transmitters within a certain distance to ensure that this is the case. Further, it must be ascertained that the new link does not cause unacceptable threshold degradation to existing users. It follows that a database in which to keep records of all microwave links is needed so that mutual interference can be assessed whenever an application for a new link is received. Additionally, if a policy similar to that described here is adopted, users will be told at what power to transmit rather than simply being given a 'yes/ no' answer to an application for a microwave link.

The above description is an over-simplification. Further questions that must be asked are the following.

- How is the total interference between different sources to be aggregated?
- Noting that the interference power will be time-varying, for what time percentage should the calculation of threshold degradation be undertaken?
- Should a separate interference-level limit be adopted when ducting is experienced?

The above questions are considered by regulators but have proven very difficult to answer in a rigorous manner. Generally speaking, where there is uncertainty, a cautious approach is adopted.

Link design at 23 GHz

As a second exercise we shall consider a microwave link at a higher frequency, such that fading due to rain attenuation is the major consideration. It is generally not feasible to provide microwave communications via a single link over distances such as 40 km when significant rain attenuation occurs. However, where the required distance is smaller, the higher frequencies permit smaller antennas to be used. Further, the use of higher frequencies where they can provide links of sufficient reliability eases congestion of the spectrum at the lower frequencies. Licensing authorities often insist on higher microwave frequencies being used for shorter-distance links. As an example, let us consider a link of a similar capacity (16 Mbit/s) to before being established over a distance of 15 km using a frequency of 23 GHz. Rain attenuation is the dominant factor at this frequency (if the peak rain rate is significant, say over 20 mm/h). In order to minimise rain attenuation, we shall use vertical polarisation.

Assuming a peak rain rate of 40 mm/h, the attenuation at 23 GHz is approximately 4.2 dB/km for vertical polarisation (see figure 7.1). However, we need to find the effective path length (the principle being that it will not be raining at the peak rate for the entire 15 km of the path). Recommendation ITU-R P. 530 suggests the following method for determining the attenuation along a 15-km path for a rain rate of 40 mm/h:

$$d_{\text{eff}} = \frac{d}{1 + d/d_0}, \tag{7.12}$$

where

$$d_0 = 35e^{-0.015R_{0.01}}, \tag{7.13}$$

$R_{0.01}$ being the rain rate exceeded for 0.01% of the time. Thus, for a rain rate of 40 mm/h,

$$d_0 = 35e^{-0.6} = 19.2 \tag{7.14}$$

and therefore

$$d_{\text{eff}} = \frac{15}{1 + 15/19.2} = 8.4 \text{ km.} \tag{7.15}$$

Thus the attenuation at the 0.01% rain rate of 40 mm/h can be predicted to be 8.4×4.2 = 35.3 dB. The required capacity of the link is 16 Mbit/s as in the previous example. Therefore the minimum required signal level is expected to be −82 dBm. In order to achieve the required availability, the link must be designed to operate satisfactorily with a fade of 35.3 dB. Thus the unfaded signal level must be designed to be −46.7 dBm.

For a link of length 15 km at a frequency of 23 GHz, the free-space path loss is given by 92.4 + 20 log 15 + 20 log 23 = 143.2 dB. A typical transmit power at 23 GHz is somewhat lower than at 7 GHz, perhaps 24 dBm. To achieve an unfaded receive signal level of −46.7 dBm, antenna gains totalling 72.5 dBi are required. Thus an antenna with a gain of approximately 36 dBi must be provided.

The gain of such an antenna can be estimated using the same method as used for the 7-GHz link:

$$\text{antenna gain (dBi)} \approx 18 + 20 \, \log_{10} f \, (\text{GHz}) + 20 \, \log_{10} D \, (\text{metres})$$
$$36 = 18 + 20 \, \log_{10} 23 + 20 \, \log_{10} D$$
$$D = 10^{(18-20 \, \log_{10} 23)/20} = 0.35 \, \text{m}. \tag{7.16}$$

Thus an antenna of diameter 35 cm would be appropriate. This can be compared with the diameter of 70 cm required for the 7-GHz link (for a distance of 40 km).

7.2.2 Broadcast systems

The ideal location for an antenna broadcasting radio or television signals is in an elevated position with a clear view to the horizon. Broadcasting is a one-way communication. It is necessary to make assumptions regarding the type of receiver that will be used and its location. It is generally expected that audio radio transmission can be received 'on the move' and indoors using small portable receivers. Thus the signal must be strong enough to tolerate losses experienced on entering and propagating within buildings. This affects the area that a single transmitter can cover. Television receivers are expected to have a directional receiving antenna located in an elevated position, such as on the roof of residential premises. This provides an advantage to television transmission, an advantage that is necessary because the bandwidth of transmission is several hundred times that for an audio signal.

National broadcasting networks will need to re-use frequencies in different locations. It is important that interference between these two regions does not occur. When assessing the interference level that will be caused, it is important to conduct the assessment using interference predictions of signal strength for the small percentage of time when ducting occurs.

By giving due consideration to coverage and interference predictions, the network designer will produce a coverage and frequency plan. This will include locations of transmitting stations together with details of antenna heights, gain and transmitting power and, additionally, the

Table 7.1 *Field strength produced at a distance of 80 km from transmitters using antennas of various heights*

Height (metres)	Predicted field strength for ERP of 1000 W (dBμV/m)	Transmit power required to produce 40 dBμV/m (dBW)
10	13	57
20	17	53
37.5	20	50
75	24	46
150	30	40
300	36	34
600	45	25
1200	56	14

frequency to be used by each transmitter. A well-designed network is crucial to ensuring good coverage and interference-free reception. It should be noted that the advent of digital broadcasting is changing the methodology by which networks are planned. Digital signals are far better for operating in a high-interference environment than analogue systems. This can lead to them demanding much less spectrum for the same number of transmitted channels (or to one being able to transmit a much higher number of channels in the same spectrum).

Suppose that we are required to design a system, broadcasting at a frequency of 100 MHz, that must produce an electric field strength of 40 dBμV/m over a coverage range of 80 km at the 50% level. We are then required to predict a range over which an interfering signal of 20 dBμV/m will be detected for 1% of the time. We shall assume that all paths involved are over land only. It should, however, be noted that interference over sea paths is generally at a higher level than that experienced over land paths. Recommendation P. 1546 is known as a 'path-general' model. The curves predict a signal strength on the basis of transmit power, antenna height and distance. Using the appropriate curve gives us a variety of options. At a distance of 80 km, the field strength produced by a 100-MHz broadcast transmitter transmitting an ERP of 1000 W (considering 50% time, land path) will depend on the height. Table 7.1 gives the approximate values.

The required signal strength is 40 dBμV/m. Note that the predicted field strength for an antenna of height 10 metres is only 13 dBμV/m. But this a prediction for a transmit power of 1000 W (30 dBW). Thus the necessary field strength could be provided by an antenna of this height but the transmit power would have to be increased by 27 dB to 57 dBW (500 kW). If an antenna of height 1200 metres were chosen, a transmit power of 14 dBW (25 W) would be sufficient. The combination chosen will depend on the cost and availability of high-power amplifiers and tall masts (although placing a 500-kW transmitter on a mast of height only 10 metres will entail safety issues). Let us suppose that a compromise of a transmit power of 40 dBW (10 kW) with an antenna height of 150 metres is adopted.

We now need to determine the distance over which an interfering field strength of 20 dBμV/m will be produced for 1% of the time. We must remember that we are assuming the use of a 10-kW (40-dBW) transmitter and that the graphs predict the field strength for a 30-dBW transmitter. We are therefore interested in the distance at which the graph predicts a field strength of 10 dBμV/m. The appropriate graph in ITU-R P. 1546 predicts that such a field strength would be produced at a distance of approximately 320 km. Thus a coverage range of 80 km would be provided at a median level of 40 dBμV/m, but interference at a range of 320 km would occur at a level of 20 dBμV/m for 1% of the time. Notice that this is over a land path. If the interference path were over a cold sea, the distance would increase to approximately 450 km. In practice, it is quite common to have a mixed land–sea path and interpolation between the two predictions would be carried out on the basis of the proportion of the path that is over the sea.

It is of interest to predict the interference distances if we had adopted the '10-metre, 500-kW' or '1200-metre, 25-W' options. Now 500 kW is 57 dBW, 27 dB above the reference level. We are therefore looking for a field strength of −7 dBμV/m. The distance for a land path can be read off the curves at approximately 370 km, a greater distance than for the '150-metre, 10 kW' option. Considering the antenna height of 1200 metres, 25 W is 14 dBW: 16 dB below the reference level. Thus we are looking for a value of 36 dBμV/m on the appropriate curve. This gives a predicted distance of 190 km, shorter than for the '150-metre, 10-kW' option. This demonstrates that, if the necessary coverage can be provided by a low-power transmitter at a great height, interference powers at a greater distance will be reduced, allowing greater frequency re-use. This can be expected since interference at large distances almost always involves trans-horizon radio paths. At large distances over the horizon, the path loss is not greatly dependent upon the height of the transmitting antenna. Therefore, the fact that when an antenna is higher, the transmitting power is lower leads to lower interference powers being produced at great distances.

7.2.3 Earth–space systems

The laws of physics dictate that, in order for a satellite to be 'geostationary' (that is, it appears to be stationary when viewed from Earth), it must be at a height of 36 000 km above the equator travelling east at a speed of about 3 kilometres per second. A satellite will be allocated a position (in degrees longitude) that indicates the point on Earth vertically below the satellite. The satellite launch system must place the satellite at that location and propel it in the correct direction at the correct speed. In this situation the gravitational attraction of the Earth is all that is necessary to keep the satellite in its position. In practice a satellite will drift with time from its allocated position. When this happens, it is necessary to push it back. The 'push' must come from jettisoning something (usually a gas) from the satellite: a fan will not be effective in a vacuum. Eventually the satellite's supply of fuel will be exhausted and any further drift cannot be corrected. Because of the necessity for the satellite to be above the equator at a fixed height, its longitude is all that is necessary to define the position.

Further, if directional antennas are to be used (the normal situation), the satellite must be stabilised so that the antenna is always pointing towards the same point on Earth. The number of satellites operating at a particular frequency that can be placed in space is limited. It must be possible to discriminate between satellites using the directionality of the Earth-based antenna. Television signals from satellites commonly use a frequency of 11 GHz. The desire is to use antennas for domestic reception no bigger than about 0.5 m. Such an antenna will have a beamwidth of approximately 4 degrees (±2 degrees from its principal direction). This means that satellites transmitting at this frequency must have a minimum separation of about 3 degrees to avoid interference, limiting the total number of satellites operating at this frequency to about 120.

The link budget for satellite systems is similar to those for point-to-point links, except that the link is much longer. One advantage is that, because much of the path is genuinely free space, the amount of fading suffered is much less. The fade margin is often very small and users of satellite television will notice degradation in picture quality when there is heavy cloud and rain present.

When viewed from the satellite, the Earth subtends an angle of about 17 degrees. This is the beamwidth required of an antenna on board a satellite if it is to provide 'global coverage'. It is more common for modern satellites to use 'spot' beams to provide service to particular parts of the Earth. A spot beam may have a beamwidth of only 2 degrees and will require an antenna of diameter about 1 metre at 11 GHz to provide this beam. In order to receive a useful television signal from a satellite, a signal power of about -90 dBm is required. At 11 GHz, the free-space loss over a distance of 40 000 km is $32.4 + 20 \log (11\,000) + 20 \log (40\,000) = 205$ dB. The gain of a 50-cm antenna at 11 GHz is approximately $18 + 20 \log(0.5) + 20 \log(11) = 33$ dBi and the gain of a 1-metre antenna will be about 39 dBi. Thus the net link loss assuming that a 50-cm antenna is used on Earth and a 100-cm antenna is used at the satellite is $205 - 33 - 39 = 133$ dB. Thus, in order to receive at a level of -90 dBm, a transmit power of 43 dBm, or 20 watts, will be needed. Older satellites were incapable of delivering such power and did not transmit with such directional antennas. It was therefore necessary to deploy very large receiving antennas on Earth. The earliest commercial satellites transmitted towards Earth at a frequency of around 4 GHz and Earth-station antennas as large as 30 metres across with gains greater than 60 dBi (and beamwidths as narrow as one fifth of a degree, necessitating sophisticated tracking control) were deployed.

As a second example, let us consider a television transmission from a satellite that is transmitted from a spot-beam antenna and is directed so that its main beam illuminates a circle of diameter 3000 km. Let us suppose that the transmission power is 10 watts (40 dBm) at a frequency of 12 GHz and the bandwidth is 30 MHz. The required signal-to-noise ratio is 11 dB. We need to determine the size of receiving antenna required on the Earth to receive the transmission with an acceptable signal-to-noise ratio.

The information that the main beam illuminates a circle on the Earth with a diameter of 1500 km provides us with information on the transmitting antenna. The path length is typically 37 000 km and therefore the beamwidth can be deduced to be approximately

$$\tan^{-1}(1.5/37) = 2.3°. \tag{7.17}$$

Now, knowing the frequency, we can estimate the diameter, D metres, of the antenna from

$$\text{BW (degrees)} \times f \text{ (GHz)} \times D \text{ (m)} = 22. \tag{7.18}$$

Substituting in values for the beamwidth and the frequency gives

$$2.3 \times 12 \times D \text{ (m)} = 22. \tag{7.19}$$

This suggests that the diameter will be approximately 0.8 metres. This, in turn, allows us to estimate the gain of the satellite antenna:

$$\begin{aligned} \text{gain} &\approx 18 + 20 \log f + 20 \log D \\ &= 18 + 20 \log(12) + 20 \log(0.8) = 37.6 \text{ dBi}. \end{aligned} \tag{7.20}$$

We now know the transmit power, transmit frequency and gain of the transmitting antenna. Further, we know that the path length is approximately 37 000 km and that we need a signal-to-noise ratio of 11 dB. We do not know the level of noise power and so will have to make sensible estimates. Further, we do not know an appropriate allowance that should be built in to compensate for rain fading. Regarding noise power, the system being described here is one where the Earth station will be a domestic receiver. As such, extremely high performance cannot be expected. However, the amplifiers used with modern domestic satellite television receivers have very good performance and an overall noise temperature of 200 K is typical. Estimates of fading likely to be experienced due to rain carried out in chapter 6 for a frequency of 11 GHz suggest that an allowance of 5 dB is appropriate for rain rates of up to 40 mm/h. This figure shall be used here.

Thus, at a bandwidth of 30 MHz the noise power received will be

$$1.38 \times 10^{-23} \times 200 \times 30 \times 10^6 = 8.3 \times 10^{-14} \text{ watts}$$
$$= -100.8 \text{ dBm}. \tag{7.21}$$

The received power needs to be 11 dB above this (-89.8 dBm). To compensate for rain fading, the unfaded signal level should be designed to be -84.8 dBm.

The free-space loss over a distance of 37 000 km at a frequency of 12 GHz is

$$92.4 + 20\log(12) + 20\log(37\,000) = 205.3 \text{ dB}. \tag{7.22}$$

Thus, the power received, P_r dBm, by a receiving antenna of gain G_r dBi is given by

$$P_r = 40 + 37.6 - 205.3 + G_r = G_r - 127.7. \tag{7.23}$$

Therefore

$$G_r - 127.7 = -84.8. \tag{7.24}$$

Hence

$$G_r = 42.9. \tag{7.25}$$

The diameter of such an antenna can be estimated from

$$D \approx \frac{10^{(G-18)/20}}{f} = \frac{10^{24.9/20}}{12} = 1.5 \text{ metres}. \tag{7.26}$$

Therefore, an antenna of diameter 1.5 metres would be needed to give a high-quality picture even when it is raining at a rate of 40 mm/h. If the antenna size were reduced to 90 centimetres, the picture quality would be good under clear-sky conditions but would deteriorate at times of significant rainfall.

7.2.4 In-building systems

Indoor systems tend to be complex, with the radio wave encountering many obstacles that give rise to multiple diffractions and reflections. The electrical characteristics of the materials are often only vaguely known and, to make the situation more complicated, people, furniture and even walls can be moved. Nevertheless, buildings and, in particular, high-rise office buildings contain many potential users of telecommunications systems and could provide significant revenue if a high-quality radio communication service could be delivered. Although the distances between transmitter and receiver are usually very small, the existence of multiple obstructions (such as walls and floors) means that the path loss can be quite high and difficult to predict. Recommendation ITU-R P. 1238 gives some guidance. If we consider the equation for free-space path loss,

$$\text{loss} = 32.4 + 20\log + 20\log f \text{ dB}, \tag{7.27}$$

where f is in MHz and d is in km, for in-building systems it is more usual to quote d in metres and the equation becomes

$$\text{loss} = -28.6 + 20\log d + 20\log f \text{ dB}. \tag{7.28}$$

The effect of obstacles such as walls is to make the loss increase with distance at a faster rate than it would in free space and the approximation

$$\text{loss} = -18 + 30\log d + 20\log f \text{ dB} \tag{7.29}$$

is put forward as suitable for use. Additionally, a frequency-dependent loss per floor is proposed when communication between floors is being planned. Indoor propagation is an area in which UTD-based methods are widely used to give as accurate a prediction as possible considering the exact building geometry. However, it is almost always necessary to make fine adjustments to the network design by making signal-strength measurements post-installation.

Rather than have a single transmitting antenna, it is common for coverage in buildings to be provided for by multiple antennas connected to a single transmitter via long feeders. This is known as a 'distributed antenna system'. Some building configurations, particularly where you have long corridors, or tunnels, such as in airports, do not use conventional antennas at all but, rather, use 'leaky feeders'. These are feeder cables designed to radiate energy. Because the feeder will get very close to the mobile terminal, it is much easier to predict the path loss when using a leaky feeder than it is when using an antenna located some distance away.

7.3 Summary

Knowledge gained previously has been applied to the design of a wide variety of radio system. Firstly, a point-to-point link at 7 GHz was designed. The major propagation-related element of this case involved predicting the fade margin necessary to overcome multipath fading. A second microwave link at 23 GHz was designed. Again, fading of the received signal would be expected but, at this frequency, the major cause of fading would be rain.

As well as predicting the wanted signal, the effect of interference on systems must be acknowledged. The way interference degrades the noise-floor of a receiver has been studied and the need for network management introduced.

When designing a broadcast system the levels of both wanted and interfering signals were predicted. Choices of combination of antenna height and transmitted power were investigated in order to assess the best method of providing the required coverage whilst limiting interference.

Transmission from a satellite to an Earth station over a distance of 40 000 km was studied. Antenna sizes, both on the satellite and on Earth, needed in order to be able to establish a link capable of carrying a television channel were determined. Finally, methods of predicting path loss for propagation within a building were put forward, together with methods of improving indoor radio communication.

8   Software-based tools

It is difficult to imagine being able to design and manage complex modern radio networks without the aid of computer databases and computer-based methods for assessment of coverage and interference. The recommendations of the ITU are purposefully designed to be implemented as a computer algorithm in the majority of cases. This chapter shows how such software tools can help in the design of radio networks using fixed links and public mobile networks as examples. It is acknowledged that no signal-strength-prediction method will be perfect and it is shown how knowledge of the likely error can be incorporated into the design process.

8.1 Software tools for management of fixed-links networks

As an example, consider the process by which a new microwave link can be allocated a frequency.

- Select an appropriate frequency band.
- Using ITU-R P. 530, design the link and determine the necessary transmit power to achieve a particular required availability (e.g. 99.99%).
- Then, for each channel,
  - locate all users of the same channel and adjacent channels within a given frequency and distance range;
  - assess whether incoming interference or outgoing interference exceeds predetermined limits; and
  - perform the previous step for a number of time percentages using ITU-R P. 452.

- If a channel passes the tests, offer it for assignment.
- Register the new link in a database.

All but the first step would be greatly helped by a computer-aided tool. The nature of the ITU recommendations for such purposes has deliberately been focussed on making them suitable for software implementation and software tools will have algorithms that perform the calculations stipulated in recommendations P. 530 and P. 452. Further, it is possible for the software to step through the process automatically, assessing each channel in turn until a channel is found that meets the requirements regarding interference levels between the new link and existing links. Although a modern computer can perform a calculation using, for example, recommendation P. 530 very quickly, the fact that interference calculations must be performed with many existing links over many channels means that the time taken for each individual step must be minimised in order that a new link can be assessed within a reasonable time.

One crucial part of any software prediction tool is a digital terrain map. These maps divide the land into square 'pixels' and assign a height to each pixel. Using these maps, it is possible to produce a path profile of the wanted and interference paths. In the case of microwave links, there would usually be a check to ensure that the wanted path had adequate clearance. It is common, indeed desirable, for interference paths to be obstructed by terrain features. Recommendation P. 452 uses the path profile generated by the terrain map in order to estimate the likely strength of any interfering signal.

Figure 8.1 gives an example of a digital terrain map. The number in each square (or 'pixel') indicates the height above sea level of that square. The size of the square is known as the 'resolution' of the terrain database. Typical values of pixel size are 50 m and 200 m. The smaller the size of the pixel, the more accurate the terrain profiles that are extracted from such a map, but at the expense of time taken both to extract and to process the profile in the prediction algorithm.

In order to compute the likely depth of a fade by using recommendation P. 530, parameters such as the rain rate exceeded for 0.01% of the time and the gradient of refractive index (dN_1) need to be known. These can be stored as additional layers. That is, a matrix similar to the one containing heights in figure 8.1 can be drawn, but with each square containing, for example, the rain rate in mm/h to be used when predicting the level of rain fading.

(a)

14	18	22	22	24	26	28	34	42	38	35	33	31	32	35	39	41	41	45
29	20	16	18	13	15	16	18	35	44	41	38	26	28	29	32	42	48	50
28	22	17	14	11	10	14	22	42	56	48	47	32	32	33	32	45	52	53
35	24	17	13	16	13	15	35	49	62	61	60	52	41	36	44	48	55	55
31	28	22	31	33	27	37	38	22	59	72	69	77	22	41	60	64	61	60
27	22	27	40	45	38	51	58	41	55	72	78	71	59	45	53	56	60	64
25	29	32	35	56	58	67	53	47	53	71	74	69	56	51	47	48	52	56
28	36	41	56	77	872	82	67	52	57	61	66	58	46	50	43	41	45	50
30	41	50	65	78	88	80	65	47	52	50	56	50	41	46	42	45	46	50
37	45	58	69	81	84	86	69	50	57	55	46	52	44	40	43	54	58	56
44	50	58	61	69	71	73	63	58	65	60	50	46	40	36	43	52	58	62

(b)

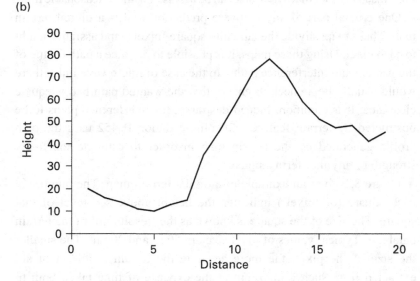

Figure 8.1 An example of a digital terrain map (a) (with heights stored for each pixel shown) with (b) a profile extracted by sampling along a line drawn on this map.

8.2 Planning tools for public mobile networks

Operators of public mobile networks are given exclusive use of particular frequencies, usually on a national basis. When planning such a network, the job of the radio-propagation engineer is to get the best out of what is called the 'air interface': the link between the base station and the mobile terminal. In doing this the engineer relies on software tools to provide estimates both of the wanted signal strength at a location and of interference (that is usually caused by other base stations within the same network). As the density of base stations increases, providing sufficient coverage is not the major challenge. Rather, the networks tend to become 'interference-limited'; that is, the benefit of adding any further base stations reduces since any new base station would both cause and suffer from substantial interference. Vital components of a software tool are

- a database to store the location and technical details of base stations;
- a map that contains terrain heights at locations with a good enough resolution (typically on a 50-metre or 20-metre grid for mobile networks);
- a map that shows land use (e.g. urban, rural etc.) at the same resolution as the terrain height map; and
- an appropriate algorithm with which to predict path loss from the base-station antennas to mobile terminals (this algorithm will consider land use at a location in estimating the path loss).

A first step would be to run the software to predict the path loss to all base stations (within a pre-defined distance) from every location within the network area. Then, for each point, the signal strength from the base station with the lowest path loss (assumed to be the wanted, or 'best', server) can be computed to form a coverage map. The engineer can inspect this map and identify any coverage 'holes' (locations where the signal from the best server is too weak to provide a useful service). These holes can be filled either by adding more base stations or by re-engineering nearby base stations. Further, the software can produce a second map showing the 'carrier-to-interference' (C/I) ratio, also known as the 'wanted-to-unwanted' (W/U) signal ratio. Areas where the C/I is low are locations where a call may fail through excessive interference or, less seriously but still non-optimum, where the capacity of the network is diminished through interference. The engineer can attempt to reduce the levels of interference, perhaps by tilting the antenna causing the interference downwards. The process continues until the network design is considered to be 'optimised'.

8.3 Mitigating against uncertainties

No propagation model is perfect. The digital terrain map itself will not be perfect, attempting as it does to sample heights at a particular grid spacing. Further, predictions of losses caused by phenomena such as rain inevitably rely on historical data that might not be a reliable forecast of future rainfall. In the case of point-to-area models used for predicting the signal strength produced by broadcasting transmitters, a single prediction is given for every pixel in the grid. In reality, this pixel could cover an area of perhaps 50 metres by 50 metres. The signal strength would vary considerably within this area. The best that any model can do is to predict the mean level within any pixel.

The network planner must examine the confidence with which any predictions are being made and make an assessment. Often the method used involves pragmatic engineering judgment as much as it involves rigorous scientific analysis. For example, within fixed links, line-of-sight clearance is required. The engineer must ask himself whether he trusts his digital terrain map sufficiently to allow it to determine whether or not a link has sufficient clearance. Usually, a site survey is insisted upon to verify this vital parameter. In the case of mobile networks, the environment can vary significantly from one city to another, making a single propagation model unlikely to be very accurate in all cases. More accurate models, based on a single general model, can be produced by making measurements from a vehicle. The results of such 'drive tests' can be used to 'tune' the model to make it more accurate in a particular area.

8.3.1 Statistical methods

In mobile communications, predicting whether or not a user at a particular location will be able to communicate with a base station is based on a link budget and a path-loss prediction. The link budget determines the maximum path loss that can be tolerated and the path-loss prediction predicts the path loss to a particular location. Except that path loss predictions from a base station predict the average path-loss to a user within a pixel. The terminology used is 'path loss exceeded at 50% of locations within a pixel'. Thus, if the predicted path loss exactly equals the maximum that can be tolerated, only 50% of users within that pixel will be able to communicate with the base station. It is normal for operators to require a greater than 50% chance of connection at the edge of coverage. Typically, 80% or 85% probability of coverage is required. In order to obtain this, it is necessary to add what is called a 'shadow-fade' margin ('shadow fading' refers to the variation in path loss as users move behind trees, buildings etc. that are not indicated on the digital terrain map). It is possible to determine an appropriate shadow-fade margin knowing the likely variation within a pixel and the percentage probability of connection required at the cell edge. Typically, 5–8 dB of margin is added to the link budget. Thus, if the link budget showed that up to 150 dB of path loss could be tolerated and it was decided that a shadow-fade margin of 6 dB was appropriate, coverage from a base station would be deemed to extend up to the point where the predicted path loss equalled 144 dB. Shadow fading usually exhibits what is known as 'log-normal' fading; that is, the distribution of path loss in dB (a 'log' scale) within a pixel gives a normal distribution.

Figure 8.2 gives an example of a normal distribution. This shows a normal distribution with a mean of zero and a standard deviation of seven. The area under the curve indicates the probability of a number from the distribution lying within any given limits.

In figure 8.2 limits are drawn at ±7 (where 7 equals the standard deviation). The area enclosed under the curve between these limits is approximately two thirds of the total area under the curve. That means that two thirds of the samples of a normal distribution will be within one standard deviation of the mean. Extending this process, suppose that we knew that the maximum path loss that could be tolerated would be 120 dB but we are not content with a 50% probability of connection at the coverage edge. We therefore must deem the edge of coverage to be at a point where the predicted path loss is less than 120 dB. Suppose that the standard deviation of the distribution of path-loss values is 7 dB. If we

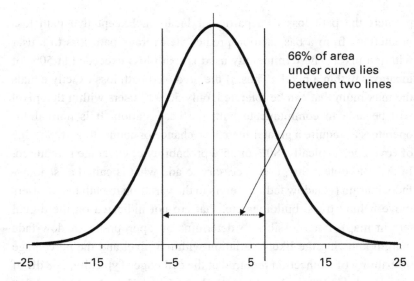

66% of area
under curve lies
between two lines

Figure 8.2 A graph showing a normal distribution with a standard deviation equal to 7.

consider now a pixel at which the predicted path loss is 113 dB (one standard deviation less than 120 dB) then only approximately 17% of samples will have a path loss greater than 120 dB. This value was obtained using the graph of figure 8.2 as a guide: 66% would have a loss between 106 dB and 120 dB; 17% would have a loss less than 106 dB; and 17% would have a loss greater than 120 dB. Thus the coverage probability, if we include one standard deviation as a margin, increases from 50% to 83%.

To reduce the necessary margin to a minimum, it is necessary to produce the most accurate path-loss model possible so as to minimise the standard deviation of error when comparing measurements against the predicted value. Operators of mobile networks will achieve this by means of tuning the model in a particular environment. Figure 8.3 shows a plot of measurements made and a best-fit line drawn through them. The standard deviation of the points about the best-fit line is 10 dB.

One possible element of the tuning process is to examine the area in which the measurements were made and attempt to define the area in such a way that the area definition could be fitted into the prediction

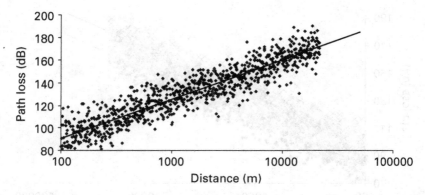

Figure 8.3 Points showing measurements together with the best-fit line. The standard deviation of the difference between the line and the points is 10 dB.

algorithm. For example, suppose that the measurements plotted in figure 8.3 were made in an area that could be divided into two categories: 'open land' and 'suburban'. The measurements could be separated into two and indicated as such on the graph. This is done in figure 8.4

It can be seen that the path loss in the suburban area is generally higher than the path loss in the area categorised as open. Most importantly, the standard deviation between the measurements and the best-fit line for each category is less than 10 dB. In this example, the standard deviation within each category is 7 dB. Thus, if the category were known, it would be possible to select the most appropriate best-fit line with which to predict the path loss and the standard deviation of error would reduce. This is done in practice. The category is often known as the 'clutter category' or 'morphological category'. Of course, it is necessary for the digital map to contain information about the clutter category and this would be another mapping layer, usually at the same resolution as the terrain map.

As an example, let us assume that we wish to estimate the coverage from a particular base station. Its height is 40 metres and the model to use in such circumstances is

$$\text{path loss} = 134 + 35 \log d \text{ (km) dB.} \tag{8.1}$$

The maximum path loss that can be tolerated is 138 dB and a coverage probability of 83% is needed at the edge of coverage. State what

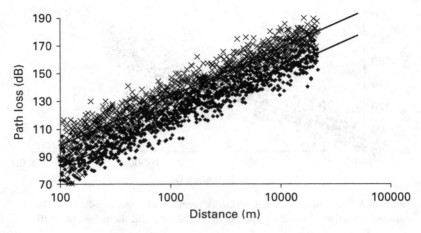

Figure 8.4 Points showing measurements made in an open area (solid dots) and in a suburban area (crosses) with a best-fit line drawn for each category.

coverage range could be confidently achieved if the standard deviation were (a) 10 dB and (b) 7 dB.

To achieve a coverage probability of 83%, we need to add a margin equal to one standard deviation. Thus, if the standard deviation were 10 dB, we would have to plan for a maximum path-loss of 128 dB. Using the path-loss equation:

$$128 = 134 + 35 \log d \text{ (km) dB,} \tag{8.2}$$

$$d = 10^{-6/35} = 0.67 \text{ km.} \tag{8.3}$$

If the standard deviation were 7 dB, then the maximum path loss planned for would be 131 dB and the coverage range would be

$$d = 10^{-3/35} = 0.82 \text{ km.} \tag{8.4}$$

The increase in range possible through use of a more accurate propagation model is significant (an increase in range of 22% that would result in a reduction in number of base stations required over a given area by 48%).

8.4 Summary

A digital map showing the heights at all locations within the area of interest is vital when making signal-strength predictions. Further location-based information, such as peak rain rate, is also required in order to predict the level of fading on a microwave link. A microwave-link-planning software tool would be able to use the terrain database to assess clearance and then use a method such as ITU-R P. 530 to predict link performance. Propagation predictions in the design of public mobile radio networks are necessary in order to evaluate the levels of coverage and interference because mature networks tend to be limited not by having insufficient base stations to provide coverage but, rather, by mutual interference between base stations. A software tool would be used such that the capacity of a network can be maximised.

It is possible for a software tool to allow for a margin for error such that the probability of making, for example, a connection to a base station was acceptably high over an area for which the tool predicted that there would be coverage. It is seen that the more accurate the propagation model, the less need for such a margin for error. Also, it is seen that making the propagation model more sophisticated (perhaps by categorising the environment) can lead to a smaller error.

9 Summary

An attempt has been made to provide an insight into the way in which signal strength can be predicted for a variety of situations. Information has been presented with the intention of stimulating an intuitive understanding of radio wave propagation together with essential formulas that will allow rapid estimates of signal strength to be made. It is the sort of information that experienced radio-propagation engineers will carry around in their heads (with the exception of the more complicated equations). Further, detailed information will be gained from consulting more detailed books such as those recommended as further reading and the ITU recommendations (also listed). Further, a radio-propagation engineer will often have software modules available that implement the ITU recommendations and other methods for propagation prediction such as the Okumura–Hata method.

Although radio wave propagation is really a single subject, all the diverse factors that affect the strength of a received signal make a comprehensive calculation of signal strength almost impossible. As a result, radio-propagation engineers concentrate on the factors that have the most significant effect for the circumstances in hand. It is seen that the task of predicting the signal received when propagation is in free space is relatively straightforward and depends upon antenna gains, path length and frequency. The concept of antennas possessing gain, although they are passive devices, is explained: the 'gain' is associated with the ability of an antenna to direct the transmitted energy in the required direction and prevent the energy spreading as it travels. Further, antennas possess the same gain whether used as transmitters or receivers. Larger antennas will direct the energy into a narrower beam than smaller antennas will, and thus the former possess higher gain. For antennas of equal size, the gain will be higher at higher frequencies: it is proportional to the square of the frequency. Antennas designed for point-to-point systems are distinctly different in nature from those used for broadcast purposes, although the same basic principles hold true. Broadcast-type antennas are often formed as arrays of smaller elements, a typical element being a half-wave dipole.

In real life radio propagation only rarely occurs in free-space conditions. One significant departure from the free-space assumption is necessary when the radio wave is in the proximity of physical objects. The radio wave will diffract around, reflect off and penetrate through such objects. Diffraction, reflection (or 'scatter') and penetration are all specialist study areas in their own right, but general rules exist that allow the radio-propagation engineer to estimate the resultant signal strength. Often, the signal produced at the receiving point is the result of multiple propagation paths including all these mechanisms. The resultant signal is the phasor summation of all possible paths. Accurate phasor summation requires knowledge of the path to within a small fraction of a wavelength. If this is not known then approximations, such as adding the power contributions for each path, are used to estimate the resultant signal strength.

Apart from the effect of physical objects, the effect of the atmosphere on radio waves as they propagate is a very significant element of a propagation engineer's work. The atmosphere is not free space, it has a refractive index that varies with temperature, pressure and humidity. Variations in this refractive index cause the radio wave to travel along a curved path rather than in a straight line. This in itself is not a problem. However, on occasion the variations in refractive index are such that the radio wave can travel from the transmitter to the receiver by more than one route. This can lead to multipath fading and is a major problem on long-distance microwave links. An additional atmospheric effect is known as 'ducting'. This can occur when a layer of high-refractive-index atmosphere becomes sandwiched in between regions of lower refractive index. The energy then becomes trapped in the layer of high refractive index and propagates large distances with relatively little loss. It is under these circumstances that interference is usually maximum. Predicting the probability and extent of multipath fading, rain fading and ducting is a duty of radio-propagation engineers. They rely on recommendations of the ITU in making such predictions.

The design and implementation of a radio system requires knowledge of a range of propagation topics to be applied at the same time. The required system must deliver the quality of service demanded of it in the presence of potential interferers. Further, the system must not cause harmful interference to existing users of the spectrum. It is common for engineers to utilise software tools in order to carry out their tasks. Such tools will have a database that stores existing transmitting and receiving stations and will have propagation algorithms embedded within the software.

Finally, it should be noted that research into radio wave propagation is very much ongoing. As spectrum becomes congested, ever higher frequencies are investigated with a view to determining how appropriate they are for particular services. Frequencies as high as infra-red are under investigation. At currently used frequencies, more accurate models are sought. These will allow any 'margin for error' to be reduced. We have seen that reducing the need for a margin for error in mobile radio planning actually reduces the number of base stations required to provide coverage. In networks such as point-to-point links, accurate prediction of interference levels and the fade margin required will allow more links to share the same frequency in a given area whilst maintaining confidence that incidences of harmful interference are acceptably low.

Appendix 1 The decibel scale

Radio engineers almost universally express powers, gains and losses in terms involving decibels (dB). This is a logarithmic scale. Converting to logarithms gives the advantage that it is possible to achieve the equivalent of multiplication by means of simple addition. Radio engineers actually 'think' in terms of decibels, which means that there is no need to convert back into normal 'linear' units at the end of any calculation: the result in logarithmic units is perfectly normal. Since calculating the effect of losses and gains in systems involves repeated multiplication and division, operating in logarithmic units reduces this to addition and subtraction.

The fundamental rule is that a power increase by a factor of ten is said to be an increase of 1 bel. So an increase by a factor of 100 is an increase of 2 bels, a factor of 1000 is an increase of 3 bels and so on. The general rule is that

$$\text{increase in bels} = \log_{10}(\text{linear increase factor}). \qquad \text{(A1.1)}$$

The bel as a unit came to be thought of as too large and power increases are now expressed in tenths of a bel, or 'decibels'. The symbol for decibel is dB: the upper case B indicates that it is an abbreviation of Bell in honour of Alexander Graham Bell (this is analogous to mW for milliwatts). This leads to the relationship

$$\text{increase in dB} = 10 \log_{10}(\text{linear increase}). \qquad \text{(A1.2)}$$

Thus a power amplifier that has a gain of 100 (and would give an output power of 100 watts if the input power were 1 watt) would have a gain of 20 dB.

Losses can be expressed in dB. If an attenuator gives an output power of 1 watt when the input power is 100 watts, the attenuation is said to be 20 dB. In the same way as an attenuator with a loss, in linear units, of 100 can be said to have a 'gain' of 1/100, so the attenuator can be said to have a 'gain' of −20 dB.

The real power of the decibel scale becomes apparent when, as well as using dB to denote gains and losses, actual power levels are described on the decibel scale. To do this, the power is expressed in dB relative to some reference level. The milliwatt is a commonly adopted reference level. Powers referenced to milliwatts are given the unit dBm:

$$\text{power in dBm} = 10 \log_{10}(\text{power in milliwatts}). \qquad (A1.3)$$

Thus $100 \, \text{mW} = 20 \, \text{dBm}$, $1 \, \text{mW} = 0 \, \text{dBm}$ and $1 \, \mu\text{W} = 0.001 \, \text{mW} = -30 \, \text{dBm}$.

The decibel scale is a somewhat more pleasing method of dealing with a situation in which massively different powers are common. This is the case in radio transmission. A transmitter may transmit as much as 1×10^6 watts but receivers can detect signals as weak as perhaps 2×10^{-14} watts. These powers would be expressed as 90 dBm and −107 dBm, respectively.

Combining absolute power levels in dBm with gains or losses in dB can cause confusion.

Suppose that an amplifier has a gain of 100 and the power at the input is 10 mW. The output power would be 1000 mW. An equation that summarises the gain provided by the amplifier is

$$10 \text{ mW} \times 100 = 1000 \text{ mW}; \qquad (A1.4)$$

10 mW equals 10 dBm, 1000 mW equals 30 dBm and a gain of 100 is a gain of 20 dB. Therefore, in logarithmic units, the equivalent equation is

$$10 \text{ dBm} + 20 \text{ dB} = 30 \text{ dBm}. \qquad (A1.5)$$

The above equation is correct. It may appear odd, since we seem to be mixing dBm and dB, but the +20 dB is identical to a ×100 operation. You may sometimes see equations like 10 dBm + 20 dBm = 30 dBm. This is quite wrong. The unit dBm refers to an absolute power level. Gains and losses are expressed in dB.

To convert from dBm to the power in milliwatts is a bit more complicated:

$$\text{power in milliwatts} = 10^{(\text{power in dBm})/10}. \qquad (A1.6)$$

Apart from the milliwatt, the watt is another common reference power. Powers referenced to the watt are given the unit dBW. So 0 dBW equals 30 dBm.

Various suffixes are added to the dB symbol. Gains of antennas can be expressed relative to an isotropic antenna (dB_i) or relative to a half-wave dipole (dB_d). If it is required to multiply by the bandwidth of a receiving system, the bandwidth can be expressed relative to 1 Hz (dB_{Hz}).

Appendix 2 Phasor arithmetic and phasor diagrams

If two, or more, 'coherent' (of exactly the same frequency) sine waves with a phase offset are added together, the result is a further sine wave of that frequency. Figure A2.1 shows an example with two sine waves with a phase difference of 90 degrees. The sum is shown as a dashed line.

The amplitudes of the two sine waves are 6 units and 8 units, respectively, and the amplitude of the sum (the 'resultant') is 10 units. The phase of the resultant is between the phases of the two components. In fact, it is nearer in phase to the larger of the two components. Rather than go to the trouble of drawing sine waves each time such a summation is required, a phasor-diagram provides a quicker method of achieving the required result. Figure A2.2 shows the phasor-diagram equivalent of the sine waves. Notice that it is possible to determine the magnitude of the resultant by completing a parallelogram using the two components.

Figure A2.2 shows a simple example of how the resultant may be determined graphically. It is possible to compute the resultant using complex arithmetic.

The above example in which there is a sine wave of amplitude 6 units that we are considering to be at the reference phase (zero degrees) and another of amplitude 8 units at a phase of 90 degrees has a resultant, R, that can be determined as

$$
\begin{aligned}
R &= 6(\cos 0 + \mathrm{j}\sin 0) + 8(\cos 90 + \mathrm{j}\sin 90) \\
&= 6 + \mathrm{j}8 \\
&= \sqrt{6^2 + 8^2}\langle\tan^{-1}(8/6) \\
&= 10\langle 53°.
\end{aligned}
\tag{A2.1}
$$

That is, the resultant has an amplitude of 10 units and a phase of 53 degrees relative to the component that we defined as the reference phase.

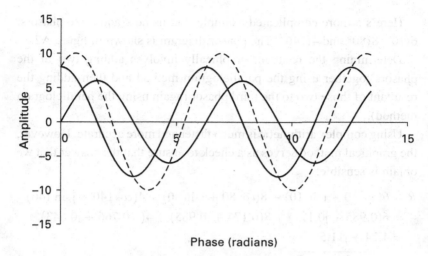

Figure A2.1 Two coherent sinusoids (solid lines) and their sum (dashed line).

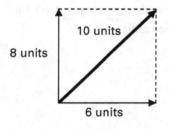

Figure A2.2 Two phasors in quadrature and their phasor sum.

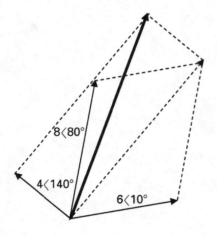

Figure A2.3 Three phasors (narrow solid lines) and their phasor sum (thick solid line). The dashed lines indicate the method of computing the phasor sum.

Here's a more complicated example. Let us now add three phasors: $6\langle10°$, $8\langle80°$ and $4\langle140°$. The phasor diagram is shown in figure A2.3.

Determining the resultant graphically involves adding two of the phasors together using the parallelogram method and then adding the resultant of those two to the third phasor (again using the parallelogram method).

Using complex arithmetic is much faster and more accurate. However, the graphical method serves as a check to ensure that the answer that we obtain is sensible:

$$R = 6(\cos 10 + j \sin 10) + 8(\cos 80 + j \sin 80) + 4(\cos 140 + j \sin 140)$$
$$= 6(0.985 + j0.174) + 8(0.174 + j0.985) + 4(-0.766 + j0.643)$$
$$= 4.24 + j11.5$$
$$= \sqrt{4.24^2 + 11.5^2}\langle\tan^{-1}(11.5/4.24)$$
$$= 12.3\langle70°. \tag{A2.2}$$

The calculated resultant, $12.3\langle70°$, agrees with that obtained by the graphical technique, giving confidence in the result.

Appendix 3   Formula sheet

Power received by an antenna of effective aperture A_{er} at distance r when a transmitter is transmitting power P_t through an antenna of gain G_t:

$$P_r = \frac{P_t G_t A_{er}}{4\pi r^2}. \tag{A3.1}$$

Effective aperture of an isotropic antenna at wavelength λ:

$$A_{ei} = \frac{\lambda^2}{4\pi}, \tag{A3.2}$$

$$A_{ei} = \frac{300^2}{4\pi f^2} = \frac{7160}{f^2}, \tag{A3.3}$$

where f is the frequency in MHz.

Antenna gain, in dBi, of an antenna of diameter D metres:

$$\text{antenna gain (dBi)} \approx 18 + 20\log_{10} f \text{ (GHz)}$$
$$+ 20\log_{10} D \text{ (metres)}. \tag{A3.4}$$

Gain of an antenna (in linear units) with a radiation pattern that exhibits circular symmetry as a function of beamwidth, θ, in radians:

$$\text{gain} = 16/\theta^2. \tag{A3.5}$$

Relationship between wavelength and frequency:

$$\lambda = \frac{300}{f \text{ (MHz)}} \text{ metres}. \tag{A3.6}$$

Effective aperture of a circular aperture antenna of diameter D:

$$A_e = \eta \frac{\pi D^2}{4}, \tag{A3.7}$$

where η is the efficiency and is typically about 0.6.

Free-space loss between two isotropic antennas:

$$L_{fs}\,(\text{dB}) = 92.4 + 20\log_{10}f\,(\text{GHz}) + 20\log_{10}d\,(\text{km})\,\text{dB.} \qquad (\text{A3.8})$$

Near-field distance of an antenna whose gain in dBi is known:

$$\text{near-field distance} \approx \frac{100 \times 10^{\text{gain}/10}}{f\,(\text{MHz})}\ \text{metres.} \qquad (\text{A3.9})$$

Near-field distance of a circular aperture antenna whose diameter, D, is known:

$$\text{near-field distance} \approx \frac{f\,(\text{MHz})D^2}{150}. \qquad (\text{A3.10})$$

Electric field strength produced at a distance d kilometres by a transmitter of power P_t watts transmitting through an antenna of gain G_t (linear units, relative to an isotropic antenna) assuming free-space propagation:

$$E\,(\mu\text{V/m}) = 5480\frac{\sqrt{P_t G_t}}{d}. \qquad (\text{A3.11})$$

Electric field strength in dBμV/m at a distance d kilometres when the product of the transmitting power and the antenna gain equals 1640 (ERP equals 1 kW) assuming free-space propagation:

$$E_{\text{dBμV/m}} = 20\log\left(\frac{222\,000}{d}\right) = 106.9 - 20\log d. \qquad (\text{A3.12})$$

Power received by an isotropic antenna as a function of electric field strength and frequency:

$$P_r\,(\text{dBm}) = E\,(\text{dBμV/m}) - 20\log f\,(\text{MHz}) - 77.2. \qquad (\text{A3.13})$$

The Okumura–Hata model for an urban environment with a receiving antenna at a height of 1.5 metres operating at a frequency of 900 MHz:

$$\text{loss} = 146.8 - 13.82\log h + (44.9 - 6.55\log h)\log d\ \text{dB}, \qquad (\text{A3.14})$$

where h is the height of the transmitting antenna in metres and d is the path length in kilometres.

Simplification of the Okumura–Hata model for a transmitting antenna of height 30 metres operating at a frequency of 900 MHz:

$$\text{loss} = 126.4 + 35.2 \log d \text{ (km) dB.} \qquad (A3.15)$$

Simplification of the Okumura–Hata model for a transmitting antenna of height 30 metres operating at a frequency of 1800 MHz:

$$\text{loss} = 136.9 + 35.2 \log d \text{ (km) dB.} \qquad (A3.16)$$

Additional diffraction loss for a knife-edge obstacle as a function of the Fresnel parameter, v, where v is greater than -0.7:

$$\text{loss} = 6.9 + 20 \log\left(\sqrt{(v - 0.1)^2 + 1} + v - 0.1 \right) \text{ dB.} \qquad (A3.17)$$

Approximate additional diffraction loss for a knife-edge obstacle as a function of the Fresnel parameter, v, where v, is greater than 1.5:

$$\text{loss} = 13 + 20 \log v \text{ dB.} \qquad (A3.18)$$

Reflection coefficient when an electromagnetic wave travelling in free space is normally incident upon a medium with an intrinsic impedance Z:

$$\rho = \frac{Z - 120\pi}{Z + 120\pi}. \qquad (A3.19)$$

Magnitude of Earth bulge, h metres, on a link of length $d_1 + d_2$ kilometres at a point d_1 kilometres from one end:

$$h \approx \frac{d_1 d_2}{12.5} \text{ metres.} \qquad (A3.20)$$

Thermal noise power, in watts, for a noise temperature of T kelvins and a bandwidth of B Hz:

$$\text{noise power} = kTB \text{ watts,} \qquad (A3.21)$$

where k is Boltzmann's constant ($=1.38 \times 10^{-23}$ joules/kelvin).

Noise temperature T_e of an attenuator of loss L (linear units) and of physical temperature T:

$$T_e = (L - 1)T \text{ kelvins.} \qquad (A3.22)$$

Approximate minimum power required to maintain an acceptably low bit error ratio on a terrestrial link as a function of bit rate:

$$\text{power required} \approx 4.0 \times 10^{-19} \times \text{bit rate (watts)}$$
$$= -154 + 10 \ \log(\text{bit rate}) \ \text{dBm}.$$

Appendix 4   Explanation
of the link budget

A further benefit from using the decibel scale becomes apparent when dealing with what is known as the link budget. Using decibels, a power budget on a radio link becomes as straightforward as a simple financial budget. Transmit power and gains can be thought of as equivalent to income, with losses and required margins being equivalent to expenditure.

For example, suppose that we transmit with a power of 30 dBm (1 watt). There are feeder losses, antenna gains, free-space loss, absorption loss, fading margin etc. The link budget is really a method of organising these parameters so as to make the calculation of the received signal level (under conditions of maximum fade, if the fade margin is considered) as straightforward as possible. The received signal level should be sufficient to deliver an acceptably low bit error ratio. Table A4.1 gives an example of such a link budget.

Table A4.1 *An example of a link budget*

Transmit power	30 dBm
Transmitter feeder losses	4 dB
Transmitter antenna gain	21 dBi
Free-space loss	126 dB
Receiver antenna gain	21 dBi
Receiver feeder losses	6 dB
Fade margin	18 dB
Received signal level under fading conditions	**−82 dBm**

Further reading

Barclay, L. (ed.) (2003) *Propagation of Radio Waves*, 2nd edn (London: IEE). This book supports an IEE course on the subject, with experts having written individual chapters. It includes chapters on individual ITU recommendations that are generally thought of as being easier to follow than the recommendations themselves.

European Commission (1999) *COST Action 231: 'Digital Mobile Radio towards Future Generation Systems'* (Brussels: European Commission). This includes details of the Okumura–Hata and Walfisch–Ikegami models.

Kraus, J. D. and Fleisch, D. A. (1999) *Electromagnetics with Applications*, 5th edn (New York: McGraw-Hill). This deals with radio wave propagation from first principles; it has a particularly good section on reflection and penetration of radio waves.

Lempiäinen, J. and Manninen, M. (2001) *Radio Interface System Planning for GSM/GPRS/UMTS* (Dordrecht: Kluwer Academic Publishers). This book goes into significant detail applying radio-propagation knowledge to planning digital cellular mobile networks.

Raff, S. J. (1977) *Microwave System Engineering Principles* (Oxford: Pergamon). This contains an excellent explanation of the principles behind antennas and noise.

Saunders, S. R. (1999) *Antennas and Propagation for Wireless Communication Systems* (New York: Wiley). This is a very thorough text that provides a detailed explanation of the mathematical foundation behind propagation theory and antenna design.

References

1. Hata, M. (1980) Empirical formula for propagation loss in land mobile radio services. *IEEE Transactions on Vehicular Technology*, **29**, 317–325.
2. Okumura, Y., Ohmori, E., Kawano, T., and Fukuda, K. (1968) Field strength and its variability in VHF and UHF land-mobile service. *Review of the Electrical Communication Laboratory*, **16**, 825–873.
3. Bullington, K. 1947 Radio propagation at frequencies above 30 Mc/s. *Proceedings of the IRE*, **35**, 1122–1186.
4. Deygout, J. (1966) Multiple knife-edge diffraction of microwaves. *IEEE Transactions on Antennas and Propagation* **14**, 480–489.
5. Epstein, J. and Peterson, D. W. (1953) An experimental study of wave propagation at 850 Mc/s. *Proceedings of the IRE*, **41**, 595–611.
6. Craig, K. (1988) Propagation modelling in the troposphere: parabolic equation method. *Electronics Letters* **24**, 1136–1139.

Recommendations of the Radiocommunication Bureau of the International Telecommunications Union (ITU), Geneva (www.itu.int)

ITU-R P. 452: Prediction procedure for the evaluation of microwave interference between stations on the surface of the Earth at frequencies above about 0.7 GHz

ITU-R P. 453: The radio refractive index: its formula and refractivity data

ITU-R P. 530: Propagation data and prediction methods required for the design of terrestrial line-of-sight systems

ITU-R P. 676: Attenuation by atmospheric gases

ITU-R P. 837: Characteristics of precipitation for propagation modelling

ITU-R P. 838: Specific attenuation model for rain for use in prediction methods

ITU-R P. 839: Rain height model for prediction methods

ITU-R P. 1238: Propagation data and prediction methods for the planning of indoor radiocommunication systems and radio local area networks in the frequency range 900 MHz to 100 GHz

ITU-R P. 1546: Method for point-to-area predictions for terrestrial services in the frequency range 30 MHz to 3000 MHz

Author's biography

Christopher Haslett has been involved in the telecommunication industry since 1975. He started his career with Cable and Wireless, where he worked as a radio-network planner. From there he moved to the University of Glamorgan, where he worked as a senior lecturer specialising in electromagnetics and radiocommunication. In 1996 he joined Aircom International, which provides consultancy and software support to the digital cellular mobile industry. His final post with Aircom was as director of planning and optimisation and his particular interest was the UMTS air interface. In 2005, Chris became principal propagation adviser within Ofcom, the United Kingdom regulator, where he acts as a point of contact for propagation matters. Further, he leads the United Kingdom delegation to the ITU study group 3 (radio propagation). Chris has a first-class honours degree in electrical and electronic engineering, and a Ph.D. in radio wave propagation. He is a chartered engineer and member of the Institution of Engineering Technology.

Index